SAINT MARY'S COLLEGE
of California
School of Education
with special thanks to the

Corporate Advisory Council

Macy's West

for their generous contribution to the
Multicultural Literature Collection

EVOLUTION, CREATIONISM, AND OTHER MODERN MYTHS

EVOLUTION, CREATIONISM, AND OTHER MODERN MYTHS,

A CRITICAL INQUIRY

Vine Deloria Jr.

Fulcrum Publishing
Golden, Colorado

Library of Congress Cataloging-in-Publication Data

Deloria, Vine.
Evolution, creationism, and other modern myths : a critical inquiry / Vine Deloria Jr.
    p. cm.
    Includes bibliographical references and index.
    ISBN 1-55591-159-5
    1. Religion and science. 2. Natural history—Religious aspects.
    3. Evolution (Biology)—Religious aspects. 4. Creationism. I. Title.
BL262 .D37 2002
291.1'75—dc21   2002008171

Printed in Canada
0 9 8 7 6 5 4 3 2 1

Editorial: Sam Scinta, Marlene Blessing, Deb Easter, Ellen Wheat
Cover design: Kristina Kachele
Interior design: Liz Jones
Cover painting: "Solar Pond" copyright © Emmi Whitehorse,
    LewAllen Contemporary

Fulcrum Publishing
16100 Table Mountain Parkway, Suite 300
Golden, Colorado 80403
(800) 992-2908 • (303) 277-1623
www.fulcrum-books.com

CONTENTS

INTRODUCTION

A COUPLE OF YEARS AGO the Kansas State Board of Education decided to de-emphasize the teaching of evolution in its curriculum, setting off a brouhaha of no small proportions. Commenting on the case, Harvard paleontologist Stephen Jay Gould reminded us that Kansas has usually been associated with the land of Oz in our folklore and dogmatically declared evolution to be a "fact"—although his definition of a fact lacked certain logic in itself. Hordes of scientific Chicken Littles proclaimed the end of the intellectual enterprise, and school principals searched their classrooms for teachers who might be offering a critical analysis of Darwinism to minds as yet not fully shaped in beliefs approved by the scientific establishment. No matter that the bookstores were filled with volumes pointing out the flaws and frauds inherent in the present articulation of evolution.

I followed this controversy with some fascination, since many well-regarded thinkers have issued consistent and prolonged criticism of

Darwinism for decades. The astounding thing about the uproar was the knee-jerk reaction among academics, most of whom could not have spoken intelligently on evolution for five minutes and who used examples that bore no resemblance whatsoever to evolutionary theory. I concluded that evolution had become a major tenet in our civil religion and, like patriotism and other generalities, was whatever anyone wanted it to be. More to the point was the realization that almost everyone involved in the debate had picked up their knowledge of scientific theory from *The New York Times* Sunday science section, *Newsweek,* or *USA Today.* When I turned to various "authorities," they seemed to know less than I did—about their own fields, in many cases.

The fundamentalists wisely hid in their bunkers during this struggle, since it was not at all clear that advocates of intelligent design and of the anthropic principle, which are intellectual ways of describing an anti-Darwin belief in patterns and purposes in nature, would come down on their side of the equation. It became clear that in addition to the age-old perspectives of science and religion, there was a third way of looking at the data, one that comforted neither the Darwinians nor the creationists. For nearly two thousand years we have believed that our solar system, indeed the cosmos itself, was a smoothly operating mechanism and that the Earth was a special project of either mother nature or god. Then the Shoemaker-Levy 9 comet hit Jupiter, and studies of the meteor/asteroid/comet hits on our planet suggested that we live on a small bull's eye that has been frequently visited by monstrous disasters of cosmic origin.

Today we receive our scientific knowledge piecemeal from two-inch newspaper columns, and each discovery is trumpeted as affirming what we already believe, so that only minor adjustments in our worldview need be made with each item. When enough discoveries begin to accumulate, however, the implications become clear: We need a major shift in our interpretation of data, and we can no longer cling to the other ways of understanding.

If each meteor hit exterminates close to 90 percent of the living organisms on the planet, how can the Darwinian "trees of life," which are supposed to show how creatures evolved, be produced? If a tsunami can deposit strata hundreds of feet thick in a matter of days, what does that imply about the validity of the slow erosion and deposition process, which has been taught as fact for more than a century?

When the smoke clears and we make all the proper adjustments in our thinking, we will come to understand that quite possibly we are not the first humanoid species to live on this planet: that there is a rough repeating pattern in the Earth's history in which the planet is transformed and new biospheres come into existence through processes of which we have not yet dreamed. This worldview is found in the traditions of non-Western peoples, including many tribal peoples. Such beliefs, which we may have previously rejected as childish superstitions, may turn out to be our only glimpse of the real planetary past.

This view, many people tell me, represents a retreat to the past. But non-Western people did not "evolve" their beliefs; they remembered events that they survived. We have cast aside these experiences because they did not fit into a neat package that explained creation, be it YHWH or the Big Bang theory, and spent our time convincing ourselves that we are the *only* example of intelligence in the universe. Thus our present knowledge is illusory because we have excluded so much data that the anomalies now outweigh doctrinally compatible evidence.

This book sketches an outline for a new way of looking at the world. The footnotes refer primarily to newspaper articles that have announced our great triumphs, but their arrangement supports the emerging paradigm, not the old one in which they are now located. I offer no comfort to religious fundamentalists or evolutionists. The views of both are passé and represent only a quarrel within the Western belief system, not an accurate rendering of Earth history. Like my earlier book, *Red Earth, White Lies,* this book will initially be bitterly attacked by smug academics, later will appear as a supplementary reading, and finally will become a major part of some college courses.

EVOLUTION,
CREATIONISM,
AND OTHER
MODERN MYTHS

CHAPTER ONE

DO WE NEED A BEGINNING?

THE CONTINUING STRUGGLE between evolutionists and creationists, a hot political topic for the past four decades, took a new turn in the summer of 1999 when the Kansas Board of Education voted to omit the mention of evolution in its newly approved curriculum, setting off outraged cries of foul by the scientific establishment. Don Quixotes on both sides mounted their chargers and went searching for windmills. The Kansas donnybrook mirrored developments in New Mexico, Texas, Nebraska, Kentucky, Oklahoma, and other states. Indeed, it suddenly seemed as if the mid continent was in rebellion against modern science. Advocates for both sides cleverly disguised their efforts to control the discussion in the most abstract and neutral terms, thereby elevating the issue of origins to a lofty status but also raising questions about the sincerity of their positions. If we take their statements at face value, the outlines of this fascinating struggle become much clearer and considerably more exciting. Since there is no reason to doubt either side in this heated debate, the arguments, dispassionately viewed, have much to tell us.

We begin with the Kansas development. On August 17, 1999, Tom E. Willis of the Creation Science Association for Mid-America in Cleveland, Missouri, helped the Kansas Board with its new curriculum and was quoted as saying that because evolution cannot be reproduced in a laboratory, it should not be taught "as though it is the only theory believed by sane individuals."[1] Speaking for the scientific establishment was Stephen Jay Gould, Harvard paleontologist and one of the most outspoken evolutionists, who argued in *Time* magazine a week later that "evolution is as well documented as any phenomenon in science, as strongly as the Earth's revolution around the sun rather than vice versa. In this sense, we can call evolution a 'fact.' (Science does not deal in certainty, so 'fact' can only mean a proposition affirmed to such a high degree that it would be perverse to withhold one's provisional assent.)"[2] But this view is not held by the entire scientific community. Gould does not see Sir Frederick Hoyle, British physicist, and Sir Francis Crick, English scientist and Nobel Prize winner, who have articulated nonevolutionary ideas, as acting in a perverse manner.

While most columnists promptly lined up in favor of science, Dennis Byrne of the *Chicago Sun-Times,* commenting on Willis's remark the next day, introduced a note of moderation with his admonitions: "The fear, of course, is that the schools will pick the 'wrong' thing to teach, such as the idea that the universe is eternal. That was the majority view of top scientists in the 1950s, but they were wrong. Now the majority believes that the universe was created in a Big Bang—a view not at odds with religious belief. What I'm saying is that scientists also can be, and often are, wrong—and incomplete."[3] This middle position is much more responsible considering the history of science and how often the majority has not only been dead wrong but also has acted with a heavy hand to suppress minority views.

Some anecdotal evidence gathered randomly while this debate was occurring shows how both sides can and do sin against the idea of free inquiry in education. Ken Bigman, a Kansas biology teacher who helped draft the rejected curriculum, said: "Evolution is the unifying theory of biology, and now students will get such an incomplete picture."[4] But does "unifying" mean that nothing else can be discussed seriously? Witness Rodney LeVake, a biology teacher in Faribault, Minnesota, who wanted to teach "intelligent design"—the theory that because the universe is intelligible, it cannot have occurred by chance—and was forbidden to do so: "They were afraid I'd turn my class into a Sunday school class, but that was never my intention. I just didn't want to teach evolution as a dogmatic fact when there's a lot of recent evidence that points to the opposite fact."[5] Note also that a student at Central Oregon Community College filed a complaint against instructor Kevin Haley, saying that he presented evolution as "a fatally flawed theory."[6] Haley and LeVake, in view of the many recent criticisms of evolution, had every right to teach that all was not well in Darwinian circles. And they would not be perverse in doing so.

Balanced against instances where teachers sought to bring a measure of impartiality were frightening episodes forecasting what we might possibly experience if the fundamentalists/creationists were to control public school education. In Belridge, California, a public school sought to adopt a Christian textbook, and the American Civil Liberties Union rushed to protest. Cited in newspaper accounts was the idea that although American Indians "attained a degree of civilization, they had no knowledge of the true God, and without this knowledge all other attainments are worthless."[7] The question of origins seems to be caught between mindless religious propaganda and narrow, unrelenting scientific orthodoxy. Perhaps more frightening is the prospect for the future. A poll by People for the American Way found that four of five Americans support teaching creationism as well as evolution in the public schools.[8] And Gallup reported that 44 percent of Americans advocated a biblical creationist view, 40 percent held a belief in "theistic evolution," and only 10 percent were strict, secular evolutionists.[9] We can be certain, then, that the conflict will not vanish.

The controversy continued in 2000 when a committee of Kanawha County, West Virginia, science teachers rejected a previously approved textbook titled *Of Pandas and People* because it discussed "intelligent design," which some teachers on reappraisal believed was creationism in disguise. They thereby limited the teaching of origins to the Darwinian thesis, which has been under attack from many quarters.[10] The next day the Oklahoma State House of Representatives passed a bill requiring science books used in Oklahoma schools to acknowledge that "human life was created by one God of the universe."[11] In May 2000 Stephen Jay Gould, now acting as the high priest of scientific orthodoxy, told a conference at Brown University that the debate over creation was "a bizarrity," describing the movement as a local American affair. And Sarah Fogarty of the science department at Providence, Rhode Island's Lincoln School warned, "We can't dismiss creationism as being fanaticism, because it's not."[12] But no evidence to exclude creationism from the ranks of the fanatical was offered.

On the offensive to crush this creationist uprising, the American Association for the Advancement of Science commissioned a report by the Thomas B. Fordham Foundation to determine how well the states were doing in teaching evolution.[13] Surely here was a survey to find politically correct curricula. Grades of D went to Arkansas, Kentucky, Wisconsin, Virginia, Alaska, and Illinois; F's went to Wyoming, Maine, Ohio, Oklahoma, New Hampshire, Florida, North Dakota, Georgia, Mississippi, West Virginia, and Tennessee. And of course, in a fit of adolescent pique, the report gave Kansas an F-. Illinois and Kentucky were said to have offered such minuscule mention of evolution as to almost avoid it altogether.[14] In 2001, showing they had their own minds, the Alabama educators agreed to put a warning on their biology books stating that there were serious questions about the validity of evolution.

A change of membership on the Kansas State Board of Education in the 2000 election produced a proevolution majority, and in February 2001 the board approved new science standards that closely followed the evolutionary dogmas. By a seven to three vote, Kansas succumbed to the ridicule and attacks of the evolutionists and restored the Big Bang theory of the origin of the universe and some of the evolutionary concepts that had been previously omitted. Dissenting board member Steve Abrams protested that he simply wanted to ensure that good science was taught in Kansas's schools.[15] He might have cited as evidence a current controversial book, Jonathan Wells's *Icons of Evolution,* which demonstrates that in many textbooks the major concepts of evolution are deliberately misrepresented as fact. But would anyone have listened?

In February 2002 the Ohio Board of Education decided to give a full hearing to proponents of the intelligent design theory of organic origins. From newspaper reports it appeared that the board members were seriously considering the issue, although a number felt it was simply another form of creationism. It was also apparent that intelligent design has considerably more punch than the old biblical creationism. Dr. David Haury, an influential professor in science education at Ohio State University, was quoted as saying, "Intelligent Design is about how things got started. Evolution is about how they change across time." Clearly, proponents of Darwinian evolution have substantially reduced their claims of infallibility and are now reluctant to describe evolution as the predominant biological fact. They now attempt to escape the debate over origins by reducing the scope of evolution to the simple task of explaining purported body changes in organisms.[16]

Some commentators have emphasized that this latest effort to delete the theory of evolution from school curricula and replace it with intelligent design is the new strategy of creationists who were turned away from their efforts to establish "creation science" as a valid subject for public classrooms. It is therefore a political rather than a philosophical or scientific problem. We can certainly trace the political attacks on evolution back to the Scopes trial in Tennessee, made famous by the movie *Inherit the Wind*. The issue at that time was whether evolution could be taught in public schools, with creationists holding the political trump cards. The tables have certainly turned.

Evolution is now the reigning paradigm, and it is the creationists who are seeking to have the court system break the monopoly. Substituting the Bible for evolutionary textbooks would certainly run afoul of the church/state constitutional barriers, so a much different strategy is in place. Seeking to broaden the meaning of science, creationists have first sought to create a "creation science" that offers partial explanations of Earth history insofar as they agree with the Old Testament. Alternatively, the effort is to allow all accounts of origins to be taught in the classrooms, knowing that Genesis would be emphasized. In the last forty years there has been a continuing examination by the federal courts of the possible ways that science could be changed or expanded to include what, for many Americans, is a true explanation of the origins of the physical world.

The recent cases worth discussing demonstrate a progression of thought from a strict adherence to the Old Testament to an increasingly complex agenda that more resembles secular catastrophism (the theory that much geological change occurs in planet-wide destructive events often caused by extraterrestrial bodies) than religious belief. Contemporary research that supports secular criticisms of orthodox Darwinian theory may lead to curricula in which a modified form of creationism could be taught. Then evolution would have to contend with the inconsistencies and anomalies of its own statements. Let us trace the emergence of this third framework of interpretation as it has surfaced in more recent cases such as *Epperson v. Arkansas* (1968)[17], *McLean v. Arkansas Board of Education* (1982)[18], and *Edwards v. Aguillard* (1987)[19]. A brief discussion of these decisions will show that the issue should not be decided by public relations wars or litigation.

In *Epperson* the state attempted to exclude the subject of evolution from its schools, justifying its position by finding that it was injurious to the religious freedom of those who considered it antireligious. This approach was novel but did not succeed. Justice Black concurred with the Court's ruling, but warned that "unless this Court is prepared to simply write off as pure nonsense the views of those who consider evolution an antireligious doctrine, then this issue presents problems under the establishment Clause far more troublesome than are discussed in the Court's opinion."[20] Sadly, he did not develop this argument to its fullest, which would have involved the questions of whether evolutionary doctrine was not itself a secular religion and whether the Court was thereby establishing it. The creationists have not exploited this argument and instead have relied on the tactic of proving that their religious doctrines were scientific.

McLean and *Edwards* provide a fascinating glimpse into the possible factual basis of a view of the universe that would emerge if religious writings had to be expressed in secular language. An analysis of the two cases will provide us with a better understanding of how the two sides, evolutionists and creationists, see each other and how they understand this intensely argued subject. Under contention in the *McLean* case in Arkansas was the Balanced Treatment for Creation Science and Evolution Science Act, passed by the Arkansas legislature in 1981. "Creation science," as defined by that body, "means the scientific evidences for creation and inferences from those scientific evidences." Creation science, then, would be based on the scientific evidences that could be used to support a theory of the origins of the universe that was more compatible with religious teachings. It would include the following points:

1. the sudden creation of the universe, energy, and life from nothing
2. the insufficiency of mutation and natural selection in bringing about the development of all living kinds of organisms from a single organism
3. changes only within fixed limits of originally created kinds of plants and animals
4. a separate ancestry for man and apes
5. an explanation of the Earth's geology by catastrophism, including the occurrence of a worldwide flood
6. a relatively recent inception of the Earth and living organisms[21]

In contrast to these tenets of creation science, the legislature defined the beliefs that are deemed representative of evolution science and that, although not formally endorsed by scientists, are not an unfair rendering of the orthodox structure of scientific belief. "Evolution science" in the eyes of the legislature meant the scientific evidence for evolution and the inferences that could be drawn from that evidence. It would include these points:

1. the emergence by naturalistic process of the universe from disordered matter and of life from nonlife
2. the sufficiency of mutation and natural selection in bringing about the development of all living kinds of organisms from simple earlier kinds
3. the emergence by mutation and natural selection of all present kinds of plants and animals from simple earlier kinds
4. the emergence of man from a common ancestor with apes
5. an explanation of the Earth's geology and the evolutionary sequence by uniformitarianism

6. an inception several billion years ago of the Earth and somewhat later of life[22]

The court was asked to rule on whether the provisions of this statute gave unconstitutional support for "religion"—a thinly disguised way of saying "Christian fundamentalism." When the trial began, prominent scientists and theologians of both sides rushed to Little Rock to defend their beliefs and demonstrate their ignorance. One can only weep for the judge who had to choose between competing squads of experts, since his understanding of the nuances of either science or religion must be suspect. Thus errors of interpretation and emphasis abound in the decision.

The judge became confused about what had been argued by the expert witnesses. He wrote that "among the many creation epics in human history, the account of sudden creation from nothing, or *creatio ex nihilo*, and subsequent destruction of the flood is unique to Genesis,"[23] which is clearly overreaching. His statement is certainly not accurate in a comparative sense. Robert Cummings Neville, dean of the School of Theology at Boston University, in *Behind the Masks of God*, writes: "The conception of creation *ex nihilo* therefore is vague with respect to whether the creator God is to be specified in a theistic sense, or in a Buddhist or Hindu sense, or in a sense congenial to Chinese religions. Each one of those traditions can be a specification of creation *ex nihilo*, by comparing them as alternative specifications of the vague notion that we can determine whether they are contradictory, supplementary, overlapping, or incommensurable."[24]

The judge stated, "The argument that creation from nothing in 4(a)(1) [the section of the statute in question] does not involve a supernatural deity has no evidentiary or rational support,"[25] which would have pleased Neville, who noted, "The bare abstractness of this bare, ontological notion of creation *ex nihilo* cannot be exaggerated. No religion treats this notion as such, only with specified filled-in versions of it."[26] In other words, the idea of creation from nothing has no meaning in itself, and every religious tradition must place a context around the statement in order for it to have validity. Does this requirement also apply to the scientific theory that the universe began with a Big Bang? What do we mean when we say it? Are we not talking about the mathematical formulas and empirical data that we have arranged to answer the question of how things originated? Do we expect science to tell us how things got under way without asking whether or not the question is a valid one? We could easily note that our most accurate knowledge is the simple observation of empirical evidence and that positing a beginning or even a steady-state universe is moving beyond the evidence to speculation.

The judge properly noted that "the emphasis on origins as an aspect of the theory of evolution is peculiar to creationist literature," and curiously, this is true. But he continued: "Although the subject of origins of life is within the province of biology, the scientific community does not consider origins of life a part of evolutionary theory. The theory of evolution assumes the existence of life and is directed to an explanation of how life evolved."[27] This statement is patently false and illustrates the confusion existing within the popular mind regarding the beliefs and doctrines of the scientific community. Although scientists may not disclaim such a broad definition on the witness stand, one can hardly read anything produced by evolutionists that does not purport to explain the origin of life itself. Why else would Sir Francis Crick feel impelled to offer a radical hypothesis of basic life forms originating in space and being deposited here as an alternative to orthodox evolution? Why would Harold Urey, Nobel Prize winner in chemistry, attack petri dishes filled with chemicals that he and others hypothesized as the original Earth atmosphere? How are subsequent events after the Big Bang explained except through an evolutionary interpretation? No, the origins of biological life are an integral part of evolutionary doctrines.

Another good example of confused thinking is the Court's description of the clash between catastrophism and uniformitarianism. "Section 4(a)(5) refers to 'explanation of the earth's geology by catastrophism, including the occurrence of a worldwide flood.' The act is referring to the Noachian flood described in the Book of Genesis. The creationist writers concede that any kind of Genesis Flood depends upon supernatural intervention. A worldwide flood as an explanation of the world's geology is not the product of natural law, nor can its occurrence be explained by natural law."[28] Here it appears that in his interpretation of the statute the judge is reacting to fundamentalist literature with which he may be familiar, and that the fundamentalists have grossly overreached themselves in insisting on only one possible planetary flood and one possible explanation of it.

Today we have greatly expanded the horizon of our understanding about the history of our planet. Certainly the Luis Alvarez meteor[29] that killed the dinosaurs, the comet Shoemaker-Levy 9 that wreaked havoc on Jupiter in 1994, and the hundreds of current studies on meteor, comet, and asteroid "hits" on our planet suggest that some floods of worldwide significance have occurred in the past as a result of hits by objects from outer space. Minimally we must take into account the massive tidal waves caused by these encounters, which would certainly have caused significant flooding of large areas of the Earth. We cannot believe the oceans would be undisturbed by the Alvarez meteor or significant hits by other space debris. Indeed, the consensus of orthodoxy in both geology and evolutionary biology is that each geologic period ends with a massive extinction of life transcending local conditions. At least one and perhaps many more of these extinctions involved a worldwide flood, whether it is that of Noah, Atlantis, or whatever cause is personally pleasing to the writer.[30] Obviously the judge had not done his homework on this issue.

The creationists' position, represented by Arkansas, introduced mathematical evidence—"the probability of a chance chemical combination resulting in life from nonlife"—as demonstrating that they had constituted proof that life is the product of a creator. Here the judge's logic is helpful. His summary of the issues explained: "While the statistical figures may be impressive evidence against the theory of chance chemical combinations as an explanation of origins, it requires a leap of faith to interpret those figures so as to support a complex doctrine which includes a sudden creation from nothing, a worldwide flood, separate ancestry of man and ape, and a young earth."[31] Logically, disproving the chance chemical process as a good explanation of the origin of life simply eliminates that proposal from the realm of possible explanations; it does not support creationism. Adding the flood and other elements to the argument then reveals the creationist bias. Although these corollaries might indeed flow from the inability to determine the origin of organic life, they do not logically support the critique of scientific explanations of the origin of life.

The evidence introduced by both sides, the arguments made, and the decision of the Court resemble a mindless stampede toward an unknown destination. Had the state really known its science, it could have given a much better account of itself. And had the fundamentalists not linked everything together so that creation science was a thinly disguised summary of Genesis, they might have given a better account of themselves. More than anything, this case represents the inability of American intellectuals in both science and religion to deal with real issues. It is as if we were replaying the Darwinian debates a century and a half later.

Five years later, in *Edwards v. Aguillard* (1987), the U.S. Supreme Court had another opportunity to rule on this controversy. Once again a state legislature, this time Louisiana, took the tenets of creation science and wrote them into law with the provision that whenever evolution was taught, the doctrines they perceived to be its rival must also be taught. This effort was a novel twist to gain the high ground, but it was not really a step forward. The Supreme Court stated that "we agree with the Court of

Appeals' conclusion that the Act does not serve to protect academic free-dom, but has the distinctly different purpose of discrediting 'evolution' by counterbalancing its teaching at every turn with the teaching of creation-ism."[32] Relying on the *"Lemon* test,"[33] which they later abandoned, the Court admitted that "teaching a variety of scientific theories about the ori-gins of humankind to schoolchildren might be validly done with the clear secular intent of enhancing the effectiveness of science instruction."[34] That is what Rodney LeVake and Kevin Haley sought to do, and it got them into trouble, suggesting that perhaps science is a religion, as some fundamen-talists have argued.

The Court acknowledged this possibility, or at least Justice Scalia raised the issue in his dissent: "The Louisiana legislators had been told repeatedly that creation scientists were scorned by most educators and scientists, who themselves had an almost religious faith in evolution."[35] In view of Gould's statement regarding perversity, there should be no doubt that part of the motivation of the legislature was to offer some measure of protection for these dissident thinkers. Scalia, however, suffered from traditional Western myopia, stating that "since there are only two possible explanations of the origin of life, any evidence that tends to prove the theory of evolution nec-essarily tends to disprove the theory of creation science and vice versa."[36] But this conclusion logically follows only if we narrow our intellectual horizons to two possible explanations. If, as is the custom in the Western intellectual tradition, we pose questions that can be answered only in binary form—yes/no, right/wrong—then Scalia's objection holds. What would we do if another reasonable explanation were offered that solved the problem of origins? Would such a theory have a chance to prove itself to either evolutionists or creationists?

The problem with the science/religion confrontation today is that sci-ence—particularly evolution—is offered as a secular alternative to the bib-lical explanation of how things began. Science has been trying to answer an essentially religious question and seems not to have noticed that its answers will always be mere substitutes for the original concepts in the religious equations. Thus, to dethrone god as the originator of life and substitute "mother nature" or "blind chance," as science has done, is simply to remain within the original framework of inquiry and to fail to ask the proper question. When science tries to answer questions such as that of the origin of life, the answers become a religious statement. Question a scientist about the problem and you will get a defensive response as emotional as anything you would hear from a fundamentalist.

Many thinkers have described this problem, but few apologists for science have taken it seriously, and virtually no efforts to reform the framework have been undertaken. Thus Western science is totally dependent on Western religion for its worldview. Harold Booher, in his book *Origins, Icons, and Illusions,* noted that "modern science originated in the Western Judeo-Christian world rather than the pantheistic East because of a belief in a god that transcends nature and placed man in a similar kind of transcendence. This allows man to *observe* objectively truth about nature."[37] Similarly, Oxford philosopher and historian R. G. Collingwood argued that scientists derive their motivation to practice science directly from Christianity: "Take away Christian theology, and the scientist has no longer any motive for doing what inductive thought gives him permission to do. If he goes on doing it at all, that is only because he is blindly following the conventions of the professional society to which he belongs."[38] Many other thinkers echo this sentiment. It is because we believe that humans are a special creation of the deity that we have any basis for believing that we can understand the natural world, since we are supposed to stand above it. In science, this perception is called objectivity and is loudly proclaimed but impossible to achieve.

Christianity, more specifically the Old Testament, has been the source of knowledge for a major part of the history of Western civilization. We may link the Greeks and Romans to ourselves via their philosophies and organizational abilities, but the peculiarly Western flavor of our science has been derived from the assumption that Christian beliefs contain accurate secular knowledge as well as religious inspiration. Thus we have a set of absolute beliefs uncritically accepted by science that have restricted our intellectual horizons for more than a century:

Monogenesis—the idea that all life must come from one source, held to be a creator in religion, determined to be an arbitrary, unseen process in science.

Time as real and linear—derived from Christian theology and uncritically accepted by science as the uniformitarian, homogenous passage of time.

Binary thinking—derived from Aristotelian logic (either/or) and Christian missionary zeal ("those not for us are against us").

Stability of the solar system—nothing has changed in our solar system since god created it or produced our sun.

Homogeneity and interchangeability of individuals—we allege to believe that all atoms and particles are the same, and that all humans are equal—derived from Christian theology and Greek philosophy. (Read any popular or technical article on science today, and you will find these assumptions taken for granted—without the slightest hint that perhaps they are mistaken.)

With science asserting that its answers to these questions are complete and accurate, we have inherited a strange body of doctrine that has limited our understanding considerably. With Darwin's popularity, and the addition of Marxian and Freudian thought in the last century, we have created a society in which science reigns supreme, and aside from occasional minor quarrels within the scientific establishment, there is no appeal to common sense, empirical evidence, or alternative explanations. As an example of the pervasive influence of science, we need only refer to a column in *Time* magazine by Robert Wright titled "Science and Original Sin": "Evolutionary psychologists say our 'moral sentiments' do, as Darwin speculated, have an innate basis. Such impulses as compassion, empathy, generosity, gratitude, and remorse are genetically based. Strange as it may sound, these impulses, with their checks on raw selfishness, helped our ancestors."[39] This idea is, of course, utter nonsense because it is pure speculation. It demonstrates only our arrogance and the unfounded belief that science can explain everything. If Wright's explanation were true, we would be getting more generous, and that is hardly the case with American society today.

The triumph of Darwinism meant a rejection of catastrophism in favor of the unobserved processes of uniformitarianism, since catastrophism was believed to give comfort to creationists. We used to believe that we lived in a clockwork Newtonian solar system, and the fundamentalists allowed one major catastrophe—Noah's flood—and attributed it to god's intervention. The new uniformitarianism inspired by Lyell and Darwin eliminated Noah's flood and reduced geological processes to mere trickles of change over innumerable spans of time, precluding the possibility of rare but rigorous events. The new concept of minute change over eons of time fitted perfectly with the social goals of the economic elite who ruled the industrial nations. Survival of the fittest, the popularization of a Darwinian concept, became a means of justifying social piracy. Anything hinting of disturbance in science also became a threat to the political status quo. Uniformitarian science was acceptable socially and politically even if it did not make sense scientifically. "Punctuated equilibrium"—the modern version of evolutionary change that suggests a biological development of fits and starts instead of a gradual change involving minute differences in body structure—may be rejected using the same logic: it may justify rebellions and revolutions.[40]

Within the basic Christian framework of interpretation, therefore, the real struggle has been a battle over the meaning of time. The Darwinian "species" need immense amounts of time during which we assume that small land mammals "evolved" into giant whales and adopted an oceanic habitat or apes quickly changed hand and feet structures, grew larger brains, and left jungle trees for Manhattan apartments. Endless time was used to explain the strange geological formations found on our planet as the result of hundreds of millions of years of erosion. Some geologists do continue to talk about "revolutions" in mountain building, but they are described as having occurred far, far away in another age and are therefore harmless to evolutionary theory. The glory of evolution has been that each generation of people is always (by definition) the smartest and most advanced group of humanoids ever. With god banished from the scene, we need a cosmic reason to feel good about ourselves.

In the fight to install a thinly disguised biblical view of origins, the fundamentalists are unwittingly promoting secular catastrophism as a potential contender within the realm of scientific doctrines, although they refuse to consider the theory in a religiously neutral context. Strangely, catastrophism is making giant strides within science to regain its primacy and may eventually become the new paradigm if freed from its relationship with reactionary religion. Let us then look at the characteristics of creation science and see to what degree these ideas, if put into a secular setting, could act as a corrective influence on present-day science and religion.

Sudden creation of the universe, energy, and life from nothing. Some thinkers wishing to reconcile science and religion will jump at the chance to say that the Big Bang theory is precisely what Genesis describes. However, we must remember that the Big Bang, as our current speculative explanation of the beginning of the universe, is merely the reconciliation of various mathematical formulas. It can easily be overtaken or reformulated as more data becomes available—or as our mathematics and computers become more sophisticated. Our inquiries often upset existing dogmas. Einstein's thinking, after all, was substantially modified later by Bell's theorem.[41]

Insufficiency of mutation and natural selection in bringing about the development of all living kinds of organisms from a single organism. The combination of mutation and natural selection as an explanation of the origin and development of life is the primary article of faith among evolutionists, and is held more tenaciously than most religious doctrines. Although there is ample evidence that mutations as a rule are detrimental, it is doubtful that scientists will surrender this last bulwark. But the development of other aspects of science may eventually force a quiet surrender in due time—another admission that "we knew all along they didn't work."

Changes only within fixed limits of originally created kinds of plants and animals. Here we find evolutionary science most vulnerable, for no significant changes can be found. In fact, eliminate the word "created" and replace it with "evolved," and you have the celebrated "punctuated equilibrium"[42] of Steven Stanley, Stephen Jay Gould, and Niles Eldredge. In fact, Eldredge, a paleontologist at the American Museum of Natural History, may be the creationist's best friend. In *Time Frames*, he provides both the data and the argument necessary to demonstrate the intimacy of punctuation and creation: "Punctuated equilibrium still strikes me as an exceedingly simple idea: at base it says that once a species evolves, it will usually not undergo great change as it continues its existence—contrary to prevailing expectation that indeed does go back to Darwin (and even beyond)."[43] How this statement is different from the creationist contention that species were created and did not thereafter experience significant change escapes me. The empirical evidence is the same; the difference is one of vocabulary.

This evidence easily falls within the definitions of both the Louisiana and Arkansas statutes. Robert Bakker, maverick paleontologist from

Colorado, echoes Eldredge's sentiment: "As a general rule, most species change very little from the time they first appear until their final extinction."[44] But Eldredge even suggests that adaptation to environment is not important: "It seems to be a general rule of ecological life: as long as they can find it, organisms will occupy their accustomed habitat. And since movement of habitats seems also the rule as a response to large-scale shifts in temperature, ice and seaways, we have a built-in explanation of statis: *there is now more than ever good reason to expect organisms not to exhibit evolutionary change even in the face of serious environmental modifications.*"[45] (Emphasis added.) Eldredge is honest about interpreting the data, but the data does not support the evolutionary scenario at all.

Evolution, it seems, covers a multitude of scientific sins but does not really explain anything. Species appear; they prosper in the same basic form until they disappear from the fossil record. Exactly *how* does this differ from creationist beliefs other than the two words "evolution" and "creation"? The *data* is the same—the fossil record. Why then, we must ask ourselves, do "scientists" froth at the mouth whenever someone suggests that evolution is not adequate? Do scientists have private and public beliefs? Apparently so. In fact, Eldredge confesses, "We have proffered a collective tacit acceptance of the story of gradual adaptive change, a story that strengthened and became even more entrenched as the synthesis took hold. *We paleontologists have said that the history of life supports that interpretation, all the while really knowing that it does not.*"[46] (Emphasis added.) Would that the judges in the two creationist cases had read Eldredge's confessions. Is there a perjury charge against the paleontologists here somewhere?

Separate ancestry for man and apes. Again this point calls into question the orthodox paleontological enterprise. We have scholars roaming Africa, Australia, the Middle East, and other locales in search of jawbones, femurs, and teeth whereby they can construct members of the primate family tree. But every fossil found is assigned to other branches of the hominid family tree than ours. We still have no direct convincing proof that men and apes are related in some specific evolutionary way. As is the custom with scientists, these new creatures are always "cousins," never progenitors. In fact, the chromosome difference between our species and the rest of the primates should have resolved the problem of relationships decades ago. We differ from them in the number of chromosomes and there is no way, under present theories, that we could be genetically related. We simply have a similar body structure.

Explanation of the Earth's geology by catastrophism. One of the ironies of science is the fact that Immanuel Velikovsky's idea of ancient catastrophes,[47] which was derided in a reactionary manner when he first proposed it, is now starting to be embraced. Once the comet Shoemaker-Levy 9 disrupted the belief in the tranquility of our solar system, dozens of scholars rushed to identify and study asteroid/meteor/comet hits, some of which may well have caused the major biological extinctions. Instead of the placid, clocklike, solar-system science accepted in the 1950s, today's solar system appears to be a shooting gallery in which our planet has been subject to unanticipated collisions with outer space debris. At least three scholars have offered new scenarios for Noah's flood using scientific evidence, folklore, and geology in true interdisciplinary fashion. There may eventually be more studies of this kind. What will happen when scientists themselves verify Noah's flood?[48]

A relatively recent inception of the Earth and living organisms. Here the fundamentalists have been blinded by dogma. We have no evidence that the creatures represented in the various geological strata were ever contemporaries, as would be required if the Bible's version of the origin of species were true. All we have are geologic eras in which very distinctive but unified biotic systems are found. Apart from the so-called living fossils, however, there is nothing but doctrinal belief to link the various biotic systems in an evolutionary chain of being. Once we factor sudden catastrophe into our geological timescale, we find ourselves without a reliable timeline, and neither creationists nor evolutionists can claim primacy for their views. The struggle between fundamentalists and scientists today is somewhat akin to the debates of the 1850s. But there is a radical difference in that we have considerably more evidence on which to base claims and develop theories. After more than a century of searching for irrefutable proof of evolution, the cupboard is as bare as it was when Darwin first advanced his ideas.

In February 2001 the University of New Mexico sponsored a unique event. Phillip Johnson, a critic of evolutionary theory, and Dave Thomas, an evolutionist, gave lectures articulating the two positions. Sadly, their presentations demonstrated that the debate is still being waged from a nineteenth-century perspective. According to newspaper reports, Johnson continually referred to god as the designer, but Thomas was no better, citing minuscule changes in existing organisms as evidence of the grand scheme of incremental development. Thomas also cited the old Darwinian finch story in which birds having slight differences in their beaks were understood by Darwin as evidence of evolution. Thomas failed to note that more recent studies have shown that the finches' beaks change with the amount of rainfall each year and that no evolutionary change is taking place.

Perhaps the biggest problem with current debates on this question is that while scientific data has increased almost geometrically in volume, many scientists uncritically accept Darwinism as the only way of interpreting this data. Since the fundamentalists insist on their literal interpretation of the Old Testament, we have not progressed at all from the original debate about how the world came to be. It may be possible to formulate a new understanding of the world that is not Darwinian, but to do so we must move from these pointless confrontations and let the data speak for itself. We already have a massive amount of data on how things act. Do we need to have a story on how they became what they are? Deep down, since we have no way of knowing, could we not simply admit that the question itself is impossible and invalid?

In a strange turn of events, the Supreme Court in January 2002 rejected a writ of certiorari filed by the Tangipahoa Parish, Louisiana, Board of Education, seeking to overturn a denial of a request for a rehearing on a Fifth Circuit decision invalidating a resolution of theirs that admonished the students to form their own opinions concerning the origin of life. Justice Scalia, in a stinging dissent, remarked: "Today we permit a Court of Appeals to push the much beloved secular legend of the Monkey Trial one step further. We stand by in silence while a deeply divided Fifth Circuit bars a school district from even suggesting to students that other theories besides evolution—including, but not limited to, the biblical theory of creation—are worthy of their consideration."[49]

Scalia's comment, phrased in unusually abrasive language, may be a point of view that we will see more frequently in the future. Perhaps the idea of freedom of thought for the individual will begin to achieve equality with the demands of the educational establishment that one and only one explanation for the origin of life can be believed. The proper place to settle these issues would be in free and open discussion. Do we have thinkers on both sides that could play this role? Would we not be hard-pressed to find people who could see and respond to the larger questions? Pending a move forward to open discussion, we are left with the question: Do we need a beginning to make sense of the world?

CHAPTER TWO

THE NATURE OF SCIENCE

THE CONFRONTATIONS IN THE COURTROOMS of Arkansas and Louisiana were between dedicated evolutionists and creationists, neither of whom were gracious or accurate. The recent outburst by rebellious school boards and state legislatures can be understood in a slightly different light. If we believe the Gallup poll numbers, many Americans do not see a conflict between creation and evolution, or do not consider the controversy important, and wouldn't mind if both theories were taught in public schools. Thus the well-publicized combat between science and religion described by the press may be grossly overstated.

Reduced to an abstract framework, the conflict can be described as a parochial disagreement within the Western structure of knowledge regarding origins and the nature of cosmic time as these topics can be developed through the interpretation of fossils. It is certainly a death struggle between the two ways of thinking, since it is apparent that science has replaced religion as the authoritative source of our knowledge regarding the world. We even derive emotional comfort from science by accepting, without criticism, whatever is given to us in the name of science. Fundamentalism also has its emotional power because it provides a simple alternative that requires the same kind of blind obedience to authority as does science. Unfortunately, its data, at least as currently presented, answers fewer questions than science does. The conflict might be better described as a struggle between the fanatics on both sides. Many people in the hard sciences do not feel threatened by creationist efforts, and large numbers of religious people see no need to participate in the controversy.

When placed in the context of science versus religion, the quarrel involves but a tiny part of each community. But while we all know that the fundamentalists do not represent the Christian religion, let alone the non-Western religions, many of us do believe that the evolutionists represent science. What is the nature of the scientific enterprise, and how is it threatened by introducing alternative interpretations of evidence? Can we narrow the definitions so that we can examine the issue in a more objective light? What are science and religion anyway?

Science and religion are concepts derived wholly within the Western historical experience. They do not appear as separate endeavors in the worldviews or historical memories of many other cultures. They originate in the great synthesis of medieval times when, after the introduction and absorption of Aristotelian philosophy in the West, reason and revelation came to be regarded as the two equally valid ways of understanding the world, indeed of reaching god. Aristotle explained the biotic world in a more comprehensive manner than did the scriptures, and in offering more food for thought, made man's reason rather than his faith important. Western civilization needed a neutral referee on matters of purely secular knowledge, and as science expanded and demonstrated its capability of formulating explanations and interpretations of data, it simply replaced both Protestant and Catholic versions of Christianity and became our sole source of truth.

Philosopher René Descartes transformed the competition between reason and faith by providing a new and neutral context in which the world could be understood. Out of his thinking came the modern understanding of mind and matter and the subjective and objective spheres. We have remained mired in this context because it seems to reflect commonsense experiences to us. But in terms of gathering valid knowledge about the world, this bifurcation has been detrimental. In plain language, the objective statement has been considered to be neutral, and the subjective has been considered to be unreliable. This way of viewing the source of our knowledge is crazy. We do not have objective facts because when we choose to experiment or investigate something, our choice of phenomena is highly subjective. But no matter how sublime our thoughts, they are usually worthless without some objective thing to which they must refer: Everything we experience as an object invokes a subjective response from us.

Unfortunately, in the centuries since Descartes, science has proven so popular and so useful that we now tend to gauge all knowledge by scientific criteria. Thus with everything from near-death experiences, reincarnation, and psychic powers to statements about our emotional and social lives, we want scientific validation even while claiming to recognize that science can only weigh, measure, and interpret material phenomena. In the last century we have seen a rapid expansion in the number of fields that claim to be scientific. Thanks to English philosopher Herbert Spencer, we have come to believe that we can treat other organisms and ourselves scientifically, and so have created the social and behavioral sciences and a host of miscellaneous courses of study that include "science" in their nomenclature. Many of these fields that have achieved some status within the rubric of science should be suspect. There is a great difference between experimenting with a rock and experimenting with a responsive organism.

The fact that various scholars purport to use the "scientific method" in their activities should not blind us to the fact that their fields and methods are highly subjective if not wholly arbitrary.

How do we understand science when so many fields of endeavor call themselves "science" but offer few credentials except the formulation of hypotheses, testing, and drawing conclusions? Surely there must be a boundary where real science ends and mere speculation begins. In the so-called hard sciences, we have definite physical objects that we can examine. Using mathematics, we can give reasonably precise measurements of phenomena. In many of the other fields claiming to be scientific, particularly those involving human or animal behavior, almost any hypothesis takes on the mantle of serious inquiry even when it is not; there are too few good measuring devices, and too many factors discarded, to make our conclusions meaningful. We currently accept the definition that if one uses the scientific method, that is science. Everything else is speculation or that mushy part of our lives—art, religion, and humanities.

Perhaps we should avoid using the scientific method as a way of describing activities and instead seek a definition of science based upon the kinds of data that we are examining. We could break down the broad definition of science according to the kinds of activities we do and the kinds of relationships we have with the data. In that way, we will have a more sensible way of judging the nature of the scientific enterprise without suffering the need to defend activities that have simply appropriated the title of "science."

Hans Küng, the Swiss theologian, offers such a method in a much different context, but his framework is so useful that we should give his categories a hearing. Speaking of our modern problems in formulating questions, Küng says that "it may help to use the terms 'paradigm' and 'model' interchangeably, but then to distinguish between Macro, Meso and Micro models (paradigms)."[50]

This distinction reintroduces the empirical dimension of science because it enhances our perspective on our data. Before we adopt the pretense that all research claiming to be scientific is such because we use a certain method, we should examine how we approach the natural world and what limitations we find imposed upon ourselves at the very beginning. So before we choose the appropriate models and paradigms, let us ask: What is the nature of our ability to understand the natural world? What are the possibilities and limitations of these three levels of activities?

Micro models or paradigms. The most spectacular success of Western science has been in physics, particularly the work done at atomic and subatomic levels. More recently, work with DNA and RNA has expanded the scope of our knowledge of life considerably, and in term of organisms, working at this level of minute existence has parallels with subatomic experiments. Our instruments enable us to consider phenomena that previous scientists could not conceive because they had no access to the information.

From the ability to examine virtually invisible activities have come the quantum wave, indeterminacy, and a host of other concepts that describe phenomena mathematically and suggest further experiments that promise an expansion or refinement of existing theory. At this level, the commands of Francis Bacon to conquer nature and force its secrets from it have certainly been heeded. We force the atom and its subparticles to behave in ways they might not ordinarily do. Nature is helpless before our relentless probing at this level, and we are truly the masters of our fate.

At the micro level we can say with a great deal of confidence that time, space, and matter are always internal to the experiment and may not have any substance at all. We have learned to talk of fields, and while we have admitted in the previous chapter that we are really talking about models that can be described in mathematical language and are not regarded as ultimate "reality," we nevertheless recognize the practical fact that these models and their formulas work within the scientific enterprise. We generally have a minimum of subjectivity, because at this level we are wholly dependent on our instruments to measure things, and measurement is a wholly objective process.

Macro models or paradigms. The macro scale is composed of all those areas that represent entities much larger than us. We are measuring the vast distances of space, the heat and light of stars, and the orbits and velocities of entities from large clusters of galaxies down to stray meteorites. We are also examining volcanoes, planetary weather patterns, and continental plates. We have no control over the data we are observing. We cannot pose any experiment in which we can force a star, or even a stray meteorite, a flood, an earthquake, or a hurricane to behave in a manner pleasing to us. Macro-level science is thus the opposite of the subatomic level. We may again use complicated mathematics and geometries to describe our experiments, but we are restricted to observational status. We must accept what the universe gives us, and within this context we formulate questions to be asked, but there are few variables that we can manipulate.

At the micro scale, activities are so concentrated and precise that we reach a point where time and space have no meaning because of the almost simultaneous occurrence of events in infinitesimal dimensions of space. Time on the macro scale has little meaning also. All of our projects seeking to discover the age of the universe or other processes that might need a relevant timescale become irrelevant on this scale because the numbers are too large. What does it mean to talk about the "first thirty seconds" of the Big Bang? "Seconds" are merely a handy device we use to describe a small portion of the time our infinitesimally small planet revolves around an unimportant star. For all practical purposes, doesn't everything that happened in the Big Bang suggest instantaneous creation or manifestation? Why else would we say "bang"? What is a "nanosecond" except a step in a logical reasoning process? Can't we have multiple explanations of the universe because we have not yet gathered enough data to choose a single answer?

Meso models or paradigms. This large area is the most difficult to comprehend. It is that area where almost everything appears to be, well, "man-sized." That is to say, there is a broad spectrum of possible objects for science to explain at this level. Some things are a little larger than we are; lots of things are smaller. We can manipulate some things that are smaller than we are—but not all of them. Other things are so large that we can only observe them and offer minor and inconsequential changes. Rarely, as in the case of thunderstorms and forest fires, can we attempt to manipulate things larger than ourselves.

The critical element at the meso level is participation. We can be both the subject and object of our experiments. We can do genetic experiments on various plants and animals, and now, it seems, clone human beings. We can pose experiments to test how other living things respond to unique and wholly artificial situations. We can alter the social situations of our own species and pretend that we have discovered inherent or genetic tendencies to act in certain ways or hypothesize possible worldviews and values. But these kinds of experiments are pure speculations dependent almost wholly on the scientist using our unexamined assumptions about the world.

Within the meso-level world we might posit a further distinction in the data we confront. We have the so-called social or behavioral sciences, which attempt to explain the values, knowledge, and activities of organic entities—plants, animals, birds, reptiles, and humans. These "sciences" must necessarily be participatory even though we proclaim their objectivity. We simply pretend that living things can be treated as objects when in fact treating them in that manner reveals only how they will respond to an artificial situation. We sometimes claim that we were objective observers without demonstrating how a completely objective observer can exist without having a cultural context that presupposes values and perspectives.

We also have a number of subject areas laying claim to scientific status that are dependent on the concept of history or the passage of time. Many of the Earth sciences—archaeology, paleontology, geology, and some cruder forms of biology and botany—fit into this general meso area. Familiar with the passage of time in our own lives and cognizant of our memories, we project the passage of time into the data. We then use our contemporary understanding of the functioning of natural processes to fantasize about possible past events and activities. The keynote of geology is that present processes can be used to describe past events. Paleontology and archaeology also project backward, from what we know of animals and people today, to describe how we think they looked and behaved in the past.

In summary, we can identify three basic levels of phenomena in which we gather knowledge that we classify as scientific. At the atomic and subatomic level, we primarily experiment. At the macro level, we can only observe and measure. At the meso level, depending on the phenomena, we can experiment, observe, and participate. We should honestly admit that we have virtually no objectivity at the meso level and that what we do is sophisticated speculation that has no permanence. It would seem that whereas we can generally rely on results obtained at the micro and macro levels, where we have either total control or no control at all, everything we say or think on the meso level is subject to cultural blinders, professional training, and personal background. That Darwin's survival of the fittest was pleasing to middle-class Englishmen as a means of justifying their empire and that some modern evolutionists refuse to consider "punctuated evolution" and catastrophism because of their possible political conse- quences in other areas of life should confirm the utter subjectivity of our knowledge at this level.

It is ludicrous, then, to endow activities on the meso level with the same status as those on the micro and macro levels. The meso level is filled with unfounded opinions, fantasies, and misconceptions because in the last analysis it is dependent on the cultural perspective of the scientist. Even the scientific establishment, when not in court proclaiming its doctrines, admits the sins of its practitioners. To summarize the conclusions of Stephen Jay Gould in *The Mismeasure of Man*, almost all scholars trying to achieve an objective view of the races of humankind have either committed fraud to make their points, set up obviously faulty premises on which to do their observations, or were "men of their time"—a nice excuse devised by the academy to avoid the inevitable conclusion that in every generation, orthodox science was limited, biased, or simply wrong. Here postmodernism has all but foreclosed avenues of escape for the meso sciences. The theoretical frameworks of our various disciplines at the meso level depend almost exclusively on already established doctrines and paradigms.

In the two court cases, *McLean* and *Edwards*, meso science got a free ride. The rigor and accuracy found in the micro and meso sciences were attributed to meso scientific statements. Evolution was lauded, applauded, and worshipped as the *only* way we could understand the origin and presence of life on Earth. Moreover, the courts were not aware that there were dissenting voices within science but relied wholly on the authority of the orthodox scientific community, which certainly had a vested interest in the outcome. Thus the convenient and comforting but fictional framework of meso science prevailed.

What, in the mind of the courts, were the elements that constituted science? In the *McLean* case the Supreme Court offered a list of what it considered the essential elements of scientific thinking. This list is important because it offers a definition based on general scientific methodology and consequently credits meso science with the virtues of the other areas of science. Thus science, according to this court, has the following characteristics:

1. It is guided by natural law.
2. It has to be explanatory by reference to natural law.

3. It is testable against the empirical world.

4. Its conclusions are tentative—that is, they are not necessarily the final word.

5. It is falsifiable.[51]

This list is strange in that it includes terms of such vagueness that any of the items could apply equally to almost any human intellectual endeavor. Would that necessarily make the effort scientific? What on Earth could "natural law" possibly mean in this context? Could it not be anything from Aesop's fables to Einstein's relativity? Isn't this definition of science comparable to similar statements about religion in that the generality of the proposition eliminates any meaningful content? Can we examine how these criteria should be applied to the two beliefs in conflict: evolutionary theory and creation science?

Applying the Court's scientific criteria to creation science is not difficult. If there were a natural law recognized by its adherents, it would necessarily be found in the Bible. Our experience of what constituted natural law for creationists would lead us to believe that it would depend, as so much does in religion, on the mind and emotions of the reader. How could creation science be falsifiable? Perhaps only by retreating from the literal interpretation of the Bible could we meet this criterion. Are there any tentative conclusions within creation science? Of course not.

Would evolution have fared better? It is certainly not falsifiable, either intellectually or emotionally, for its advocates. Indeed, it is holy script for many of its believers. Witness the statement of the much-beloved Pierre Teilhard de Chardin, Roman Catholic priest and advocate of evolution: "Is evolution a theory, a system or a hypothesis? It is much more: *it is a general condition to which all theories, all hypotheses, all systems must bow and which they must satisfy henceforward if they are to be thinkable and true.*"[52] Clearly for Teilhard and his followers, evolution *is* natural law. Did any pope ever speak in less absolute terms? Do most modern supporters of evolution endorse this statement? You can bet they do. If we accept Teilhard's sweeping endorsement, which many scientists and theologians gladly do, evolution becomes a religion and then is disqualified under points four and five. It is no longer tentative and it is not falsifiable, which are the essential elements of scientific thinking. We could weasel our way back to some kind of intellectual respectability by seeking to qualify evolution under points one and two, arguing that evolution is guided and explained by natural law. But evolution is always based on chance occurrences that preclude patterns of development and obey no known laws. How then could natural law be supportive of evolution?

Since many people on the science side of the conflict would accept points one and three, perhaps we need to examine the idea of natural law to find some way to qualify evolution as scientific. Strangely, people outside science seem to give considerably more credence to the absolute nature of "natural law" than do people within science. Paul Davies, Australian philosopher, writes, "Laws of nature are real, objective truths about the universe," and argues, "We discover them rather than invent them."[53] Keith Ward, English theologian, agrees, suggesting, "The reason events happen in intelligible, largely predictable ways is that they act in accordance with general principles, laws of nature. The laws of nature look just as if a wise creator had selected them as the most simple and elegant principles of intelligible change. Belief in the intelligibility of nature strongly suggests the existence of a cosmic mind that can construct nature in accordance with rational laws."[54] But the intelligibility of nature more likely suggests our intelligence than the presence of a cosmic mind.

The arguments, as everyone will recognize, are merely restatements of the medieval belief that reason and revelation stand on equal footings in terms of our ability to understand the universe. Thus the pall of Christianity intrudes into our effort to understand the apparent intelligibility of the universe. Where does natural law come from? If we make it an absolute part of the universe, or science, we must trace its origins to an initial "cause" and introduce the idea of "intelligent design," which is unacceptable to science. And we have made virtually no progress.

When we turn to thinkers of a more scientific bent, we find an uncertainty and practicality that are refreshing. Richard Milton, popular science writer and critic, points out that "at the atomic level, events occur with a randomness which appears chaotic, and it is only when we observe the statistical effects that appear when billions of atomic events are aggregated that the humanly significant events emerge that science dignifies with the term 'Natural law.'"[55] Certainly this condition holds in everything we know at the micro level, or we would not have quantum mechanics at all. "Scientific laws are always of the nature of idealized 'thought experiments,'" writes Hans Schaer, Jungian psychologist. "In the real world, contingencies and individual events tend to impose their own unique characteristics on physical law so that only averaging over repeated controlled experiments can reveal tendencies."[56]

In fact, not only is natural law an expression of averaging countless numbers of unique events to formulate general principles of change and predictability, "human reason" is in no better shape. Reason, Schaer suggests, " is the expression of man's adaptation to what is normally there, and this is deposited in complexes of ideas which, organizing themselves step by step, constitute objective values. The laws of reason are thus the laws which characterize and regulate the adapted attitude, the—on an average—'right' attitude."[57]

Having placed natural law in its proper context, as a statistical truth, thereby raising questions about the status we grant scientific statements, we move to an important point in the Court's definition of science. *It is testable against the empirical world.* We know that the "hard" sciences are indeed testable against natural phenomena on the micro and macro scales because we measure, weigh, and observe things using complex and precise mathematical formulas and using complex instruments. We know almost exactly the answers we can expect. Receiving results far afield from our expectations tells us something new about the world and enhances our knowledge. In theory, the scientist whose behavior most resembles that of the machine or instrument is the most complete scientist, since subjectivity is minimal as he or she approaches machinelike participation in the experiment. When we run electric current through something infinitesimal, we get the responses necessary for us to draw conclusions. When we analyze the light spectrum of stars, again, our instruments eliminate any possibility of subjectivity; the only subjectivity at the macro level is choosing where to point the telescope.

The meso level is a complete state of confusion. Although we can measure things, the selection of how and what to measure, and the conclusions that we draw, often fit the existing paradigm rather than the evidence. Our colleagues can easily change the rules—and the results—by rephrasing our initial inquiry. A glance at the "scientifically approved" diet plans available to us should dispel the belief that we have gathered any significant kind of knowledge. Two meso sciences are the targets of the creationists: evolutionary biology and geology. The charge is that doctrines in these fields do not correspond with the empirical evidence and that the biblical explanation does so adequately. Prior to Darwin and Lyell, the Bible was believed to contain both the religious and secular explanations of the origin and functioning of the natural world. Certainly people saw the Noachian flood as consistent with most geological interpretations. The formula for explaining the biotic world was simple: There was a preflood complete creation with all possible organisms as contemporaries and then a postflood biosphere of animals who had booked passage in the ark. Occasionally there were some strange strata containing weird fossils thrown in by the devil to confuse us and test our faith. No explanation was offered for dinosaurs or megafauna.

This framework would have collapsed of its own absurdity in time as more empirical data became available. The appearance of strange fossils of gigantic birds, various dinosaurs, and larger versions of living contemporary animals spoke of worlds not remotely suspected in the biblical accounts. Thus, whether or not Darwin had articulated his ideas, the credibility of the biblical account was failing badly when evolution rose to challenge it. Speculations about a comet's impact being responsible for the flood, first articulated by William Whiston in the late seventeenth century, had already provided the acid to begin to dissolve the biblical foundations. Today when fundamentalists insist on remaining within the biblical framework, their arguments carry little weight.

Langdon Gilkey, prominent Protestant theologian, has correctly explained the process of change initiated by evolution and uniformitarian geology in this massive intellectual revolution of the last century. "The knowledge guided by science," he wrote, "*replaces* the knowledge derived from religious commitments; and modern scientific technology in its broadest sense *replaces* the benefits brought by religious devotion and practice."[58] Here Gilkey suggests that there were two replacements as a result of the rise of science and the decline of biblical literalism. The emotional energy we had once invested in religion as an absolute source of authority was uncritically transferred to science, which then became our guarantor of truth. More importantly, the replacement process began when reasonable explanations offered by secular thinkers appealed to our common sense and were successfully compared with the mysterious workings of god's will. Granting reason equal status with revelation was the key here.

Having once accepted the biblical accounts literally, we now accepted science's findings literally. Secular science then merely substituted its doctrines for the previous religious concepts and as a consequence became a prisoner of the biblical conception of linear time. Thus "nature" replaced "god"; "science says" replaced "god's will." Ultimately, the chances of evolutionary biology meeting the test of empirical verification may be no better than that of creationist theory, since both explanations follow the same logical sequence. But many scientists would deny that they have simply appropriated the biblical framework and not allowed the evidence to speak to them of a different paradigm.

Norman F. Hall and Lucia K. B. Hall, avowed materialists, in an article in *The Humanist*, give their analysis of this contemporary quarrel. "Science and religion are diametrically opposed at their deepest philosophical levels. And, because the two worldviews make claims to the same intellectual territory— that of the origin of the universe and humankind's relationship to it—conflict is inevitable."[59] But are these really two different worldviews? Or are they the same answer to one question using different words, posed within essentially the same outmoded worldview? What is the nature of this conflict at the deepest level? It is a struggle over the *interpretation* of data, since both creationists and scientists cite the same empirical data but draw radically different conclusions.

We must make a distinction between the gathering of data and the interpretation of that data. It is the essence of the scientific task to gather data, to propose experiments, to make observations, and to suggest theories of interpretation. And we certainly need general paradigms within which we can classify data and propose further experiments. The problem is that both creationists and evolutionists take their assumptions and presuppositions from the same source—uncritical acceptance of a linear idea of cosmic time. They do not let the data speak to them. For a century and a half we have seen evolutionists looking for "missing links," which must necessarily represent chunks of missing time in a linear sequence. Creationists then applaud when proposed missing links are discredited and the timescale appears to collapse. We can demonstrate how closely the thought processes of these two groups resemble each other; we merely need look at the complaints they file against each other.

Evolutionists typically level a basic criticism of the creationists. Keith Ward, a theologian but committed evolutionist, writes, "The trouble with the God hypothesis is that it seems to explain almost anything. Nothing can disconfirm it. And believers seem almost irrationally committed to it—or at least they believe it with a strength far beyond what the evidence might suggest."[60] Steven Stanley, an English punctuated evolutionist, complains: "Whatever happens is God's Choice. Putting it the other way around, God's Choice is whatever happens, and this means that a divinity can always be invoked without the possibility of challenge."[61] How, then, does evolution differ from this tendency to cite a cause that cannot be explained and which therefore terminates the debate? It doesn't! As we have seen with Teilhard de Chardin, evolution covers everything.

For a long time, scholars of varying persuasions have leveled the same charge against evolution that evolutionists level against the creationists. Half a century ago, Robert Lowie, noted anthropologist who studied the Crow Indian tribe, in his book *Primitive Religion*, wrote: "So long as Evolution remains an unanalyzed mysterious complex of ideas sanctified from boyhood, it is taboo, set apart from the operations of logical thinking. Attempts to prove it, and the very need for proof, [are] bound to shake confidence."[62] The situation has not improved in the decades since Lowie, even though we have tripled or even quadrupled the data from which we draw our conclusions. Harold Booher recently observed: "In the central role that evolution plays in the minds of so many people it certainly can be likened unto a religious one. Although frequently claimed to be a scientific theory having no connection to religion, it is unique in science in that its devotees often describe it in terms most clearly resembling reverence for a superhuman power."[63] The two courts apparently interpreted scientific arrogance as sincerity when they should have seen it as devotion.

We often hear from the fundamentalists the refrain: "I believe in the Bible!" We never know what that statement means. Indeed, we never even know if the person doing the believing actually owns a Bible. Even when people are reasonably schooled in the Bible or at least one of the Testaments, we find that they use it selectively to support everything from capital punishment to gun ownership to discounts for trading stamps. We become deeply suspicious that the whole thing is an elaborate hoax. Unfortunately, the same can be said about evolution. Harold Booher describes the selectivity quotient among evolutionists in this manner. "Since people can believe in the factual basis of *what they each see and understand to be evolution the concept comes to mean just about anything anybody wants it to mean*" (emphasis added). And he comments, "[Evolution] is incapable of proof or disproof. Each person therefore can have his or her own personalized belief in evolution."[64]

Evolutionists even follow the traditional format used by religious zealots to prove any number of things, typically to prove that Christianity is the highest religion. "Because 'evolution' means so many different things," Phillip Johnson, law professor at the University of California at Berkeley and critic of evolution, notes, "almost any example will do. The trick is always to prove one of the modest meanings of the term, and treat it as proof of the complete metaphysical system."[65] One need only read any of the exhaustive tomes written by evolutionists to verify this observation. Examine this rather confused statement by Stephen Jay Gould regarding speciation: "Our claim is not that all speciation brings paleontologically visible change, but the very different notion that, *when* such change does occur, it occurs in concert with episodes of speciation. Similarly (for the same illogic has often been advanced in this case), the fact that most peripheral isolates do not form species is no argument against the claim that speciation, when it occurs, happens in peripherally isolated populations."[66]

Now, exactly what is Gould's argument here? A casual reader will assume that he has proven that peripherally isolated populations can produce new species. But if we examine his statements carefully, we find him saying that while isolation does not necessarily produce speciation, if any speciation does occur, it happens in isolation. But does speciation occur at all? We are left wondering. The logic here is exceedingly strange because we are confronted with negative evidence to prove a positive fact. This strange chain of reasoning passes for a scientific explanation, and after reading Gould we are filled with pride that we have been given a valuable insight into the processes of evolution.

Evolutionists always assure us that scientists cannot do without their theory—it is an essential part of every experiment they undertake and every article they publish. Indeed, Langdon Gilkey rather hysterically argues that "without this thesis of a universe in process over eons of time [what was referred to as the *fact* of evolution], there is simply no modern science, no astronomy, no geophysics, no geology, biology, paleontology, botany, anthropology, agronomy or meteorology."[67] The facts are otherwise. One of the important scientific books of recent times, *Darwin's Black Box* by Michael Behe, looks at six processes in microbiology without which we would have no possibility of life of any kind. He reaches a startling conclusion: Evolution is not necessary for sophisticated modern biological work. It hardly appears at all, and certainly not as an integral part of the scientific enterprise.

Behe carefully surveyed the literature in these several subfields to see how often and to what extent scientists needed to use the concept of evolution in their work, and if they published papers that enhanced evolution as a theory or assisted in proving its validity. The results of his research are significant:

> Molecular evolution is not based on scientific authority. There is no publication in the scientific literature—in prestigious journals, specialty journals, or books—that describes how mo-

lecular evolution of any real, complex, biochemical system either did occur or even might have occurred. There are assertions that such evolution occurred, but absolutely none are supported by pertinent experiments or calculations.... [I]t can truly be said that—like the contentions that the Eagles will win the Super Bowl this year—the assertion of Darwinian molecular evolution is merely bluster....[68]

The general professional literature on the bacterial flagellum is about as rich as the literature on the cilium, with thousands of papers published on the subject over the years.... *Yet here again* the evolutionary literature is totally missing.... [N]o scientist has *ever* published a model to account for the gradual evolution of this extraordinary molecular machine....[69]

In fact none of the papers published in JME [*Journal of Medical Endocrinology*] over the entire course of its life as a journal has ever proposed a detailed model by which a complex biochemical system might have been produced in a gradual, step-by-step Darwinian fashion.[70]

Now of course the punctuated evolutionists Stephen Jay Gould, Niles Eldredge, and Steven Stanley would no doubt respond that while this state of affairs is true, that is all the more reason to support their version of "punctuated equilibrium." But that "theory," if we can call it a theory, merely states that once a species has appeared in the fossil record, it generally does not change and makes only a few minor adjustments in body structure over millions of years.

But that is precisely what creationists say. The only difference, I repeat, is that any evolution of significance always occurs "offstage" in some unknown geological strata that have eroded away or are inaccessible. Why not say the whole thing occurred in heaven, where god is busy planning a habitat for humanity? Philip Kitcher, a stalwart evolutionist, illustrates this point. "We can understand neo-Darwinism as specifying a problem-solving strategy. When we confront a case of speciation, we are to advance a particular kind of Darwinian history, one that involves a claim that the daughter species emerged from an isolated subgroup of that ancestral population."[71] The choice of how we are to explain the daughter species, particularly when we find only cousins, is left wholly to our discretion. We can say anything we want and it will become evolution. We will examine this strange logic in the next chapter in more detail. But how does this application of theory differ from Reformation church catechisms that explain events as god's will?

Completely lacking in the scientific method in many of the sciences is some criterion whereby we can establish a modern version of natural law that avoids the enforcement of beliefs. The concept should not depend on personal preference. But we have no mechanism to allow this possibility. It is all ad hoc at the moment. Ian Barbour, a professor of both theology and science at Carleton College, observed that "no rules specify whether an unexplained discrepancy between theory and experiment should be set aside as an anomaly or taken to invalidate the theory."[72] In other words, we may have a prevailing paradigm, but we have no accurate way of knowing when or whether it should apply. We have no way of deciding what data is relevant, so we simply reject or avoid data that does not fit and call it an anomaly—a fact that, incidentally, does not obey natural law or the requirements of the paradigm.

How can people know when an anomaly should be discarded or when it points to another explanation? Cases frequently come to light in which some dreadful mismatches between data and dogma are hidden behind the consensus of the scientific community or the prestige of a prominent scientist. Should we borrow from the field of law and require that a theory be established beyond a reasonable doubt? Or by a preponderance of evidence? How can the obvious message of data compete with the doctrinal consensus of the scientific community? Everything is too uncertain at the meso level to give absolute answers.

So what is the nature of science? It appears to be a reasonable effort to gather and interpret data about the world we live in. When we believe we have found enduring patterns of behavior, we tend to formulate propositions that describe what will probably happen in succeeding cases or examples of the same behavior. We can then describe this general proposition as "natural law," while remembering that it is not law in the usual sense. It is a convenient way of expressing our expectations. Objects and entities are not bound by it. If they appear to act differently, this behavior should tell us that our proposition needs modification. By most standards of logic, neither creation nor evolution meet the practical requirements for a science. Had the two courts been more familiar with science, they would have realized that the criteria used to distinguish evolution from creation as a "science" does not hold.

CHAPTER THREE

THE PRIMACY OF SCIENCE

IT IS NOT SURPRISING that the federal courts have always turned back the efforts of conservative state legislatures to mandate equal treatment in schools for creation theories. The courts' rulings, which called these legislatures' proposed "creation science" curricula a thinly disguised retelling of Old Testament events with no evident use of the scientific method, reflect reasonable concerns about church/state separation. A well-developed theory of catastrophism using contemporary scientific studies might be an effective approach in view of the recent popularity of the subject. To survive a neutrality test, then, the creationists need to broaden their perspective and adopt a neutral agenda into which the biblical scenario can fit comfortably.

The Scopes trial, conducted in Tennessee in the 1920s, discredited fundamentalism and came to symbolize the triumph of reasonable, empirical science over deliberately ignorant ancient superstitions. Ever since then, scientists have defended their new status and posed as the ultimate authorities with great vigor. No realistic alternative to Darwinism has been offered that would either provide a new interpretation of data or broaden the general intellectual horizon beyond the Western framework to include the views of Eastern and aboriginal peoples (thereby bringing a planetary perspective to the question of origins). Evolution's competition has been not secular catastrophism but ad hoc catastrophic propositions that inevitably lead to the endorsement of the biblical version of ancient events.

Suggesting more than one flood, for example, would have irritated both parties. But Noah's flood could be included under the rubric of general catastrophes of fire, flood, and wind, and that would broaden the playing field so that the scientific data collected by the creationists could be brought into evidence. Make no mistake, the creationists have compiled a body of antievolutionary data that is impressive. But they have not used their criticisms effectively, believing that if they prove evolution wrong, some kind of divine creation is the only feasible alternative. Concentrating on the problems that contemporary biological theory has in supporting Darwinian evolution and offering a new means of interpreting the data might have narrowed the issue so that the "punctuated evolutionists" would have needed to scramble to remain within the orthodox fold. One can only conclude that while the sentiments for a more balanced view may have been proper, the lack of understanding of the issues and inability to offer an alternative paradigm doomed the creationist efforts.

A deeper problem exists, however. Science is our religion today. Because it officially produces no concept of divinity, it is regarded as neutral on the question of origins. But science is firmly grounded in a crude materialism that precludes the possibility of divinity, indeed even rejects any hint of teleology—the sense of purpose and meaning—in its explanation of the world we live in. "Chance," badly defined or not defined at all, is cited as the reason we exist. Life has no greater purpose than to survive without meaning, and we are born, live, and die without knowing why. Although we pledge allegiance to various forms of religion and ethical considerations as a matter of respectability, in fact, we generally accept what science tells us about the world.

We live in a scientific culture in which religion has long since lost its authority to speak on the most pressing issues of our time. Thus the case for creationism—or even catastrophism—needs a more sophisticated structure and presentation and should seek to take the same empirical data and offer a new vision, if not of the meaning of, at least of our orientation to, the physical world. But even if opponents of evolution had unveiled a responsible critique, it is doubtful that they could have carried the day in the courts. Judges suffer from the same handicap as everyone else. Their knowledge of the issues is restricted to the smattering of data they received in their younger days, much of which has long since been discarded.

Dutch philosopher Paul Feyerabend, critic of orthodox scientific epistemology, was inspired to observe that "science is no longer a particular institution; it is now part of the basic fabric of democracy just as the Church was once a part of the basic fabric of society."[73] And he suggested that science's "social authority ... has now become so overpowering that political interference is necessary to restore a balanced development."[74] Indeed, creationists' efforts to accomplish politically what they cannot do intellectually seem to validate Feyerabend's contention that the scientific establishment is something more than an intellectual enterprise. It represents our most fundamental beliefs. Were the fundamentalists not safely within their religious cocoon, they would realize that political struggles will generally fail and that their only recourse is to offer a complete and satisfactory alternative to the present beliefs of science.

Although fundamentalists refuse to acknowledge science's supremacy, some scientists mince no words when bragging about their status, and many old-line denominational theologians meekly consent. Thus Stephen

Jay Gould, in *Rocks of Ages,* issued this pronouncement: "Now the conclusions of science must be accepted *a priori,* and religious interpretations must be finessed and adjusted to match unimpeachable results from the magisterium of natural knowledge."[75] Modesty becomes Gould. And he warns theologians that if "some contradiction seems to emerge between a well-validated scientific result and a conventional reading of scripture, then we had better reconsider our exegesis, for the natural world does not lie, but words can convey many meanings, some allegorical or metaphorical."[76] Clearly, we see here, science recognizes that it is the dominant force in our society and that our traditional religions have been reduced to mere sentiments and homilies with nothing to offer society. If a religion deals with "ultimate reality," then science is our religion.

Kenneth Woodward, columnist for *Newsweek,* used more moderate language in supporting this perspective: "Modern science presents an increasingly compelling model of how the world works to which religion, if it is to remain intellectually honest, must adjust its ideas about God."[77] Theology then exists at the sufferance of popular theories of science and must adjust itself accordingly. Langdon Gilkey, a prominent Christian theologian and a witness for the evolutionists in the Arkansas case, admitted, "Science has become for many the supreme form of knowing and the key to effective action. From this position it is, of course, merely a step to believe that science therefore represents the *only* form of knowledge."[78] Indeed, Paul Davies, a professor of mathematical physics at the University of Adelaide, Australia, and winner of the prestigious Templeton Award, enthusiastically endorses this view: "Science hasn't explained everything, but that doesn't mean it can't."[79]

One hopes that this attitude represents the high tide of scientific arrogance, and that eventually a bit of humility will creep into the discussion. As we know from previous experience with other religions, however, once an ideology has gained power, it almost always oversteps itself and sets the stage for its demise. High priests of science such as Davies and Gould should review scientific history and tone down their remarks and optimism. The establishment has almost always been wrong—even about the findings of science. Many of the heroes of science were themselves outcasts and heretics. Fellow scientists rejected Galileo long before the church made him a martyr for science. Like representatives of all former "absolute truths," representatives of science today who trumpet its infallibility stand on very shaky ground.

Why do so many otherwise intelligent people swoon when science claims to be the possessor of "absolute truth"? Paul Feyerabend says that science "reigns supreme because its practitioners are *unable to understand* and *unwilling to condone* different ideologies, because they have the *power* to enforce their wishes, and because they use this power just as their ancestors used *their* power to force Christianity on the peoples they encountered during their conquests."[80] Ian Barbour, also a winner of the Templeton Award, says: "There is no court of appeal higher than the judgment of the scientific community itself. The presence of shared values and criteria allows communication and facilitates the eventual emergence of scientific consensus."[81] So according to these thinkers, we have no remedy once science has spoken. Whatever truth is, it will be delivered to us when the scientific community has reached a consensus. But with a few nominal substitutions, doesn't this statement have a bit of the medieval inquisitorial flavor? If we substitute "church" for "scientific community," do we not have a familiar refrain echoed here?

How do we arrive at a scientific community consensus or judgment so that we can receive scientific truth? In *Science and a Free Society*, Feyerabend describes the process in realistic terms: "Unanimity is often the result of a political decision: dissenters are suppressed or remain silent to preserve the reputation of science as a source of trustworthy and almost infallible knowledge."[82] He also notes that unanimity is often the result of shared prejudices. Scientists refuse to surrender beliefs from their under-graduate days, and if they are not current with a field, they simply perpetuate errors for another generation. Certainly much can be said for this accusa-tion, and we hear it all the time in the mutterings of scholars who lack the prestigious standing to force a change in beliefs.

A clear warning against the belief that a clear consensus can be achieved in the scientific community was given by Percy Bridgman, Harvard physi-cist and philosopher, a generation ago: "The significance of consensus is limited. 'Competent persons' at any epoch means those who at that epoch have been subjected to a definite precondition, so that consensus merely means that persons who have been subjected to this conditioning react in a certain way. It does not preclude the possibility that all competent per-sons are reacting incorrectly because of some feature in contemporary cul-ture."[83] I have often found colleagues unable to give me a single reason why they believe in a scientific doctrine other than that they learned it in grad-uate school.

Are there more modern ways to reach unanimity? Apparently so. J. Richard Greenwell, cryptozoologist at the University of Arizona, in a let-ter to *Science Digest,* reported: "Recently, two members of the American Geophysical Union proposed a 'democratic solution' to the question of the dinosaurs' extinction. Whatever hypothesis received the most votes from the Union's members should, the scientists suggested, be declared the official reality."[84] Greenwell wrote, "It is sobering to realize that with the change of a few percentage points in a poll, whole menageries of new crea-tures could enter our textbooks almost overnight."[85] Or they could sneak into the textbooks one by one as "missing links" if endorsed by prominent professors in the specialty field. With a good public relations plan, then, we could quite conceivably restore the unicorn to its rightful place in the evo-lutionary chain of being.

Although these examples may be extreme, they are well inside the scope of present-day scientific decision making. Prominent scientists can insist that their favorite doctrines be accepted as scientific truth whether or not the data supports their contention. Thus in anthropology we have had a Clovis Man inhabiting this continent in early times although the evidence for him can be mailed with a single first-class postage stamp.[86] Many doctrines are modeled after the scientist's own personal view of reality or even after his prejudiced preferences. W. B. Rouse, a systems engineer who seeks to apply methods of business management to science, said it well: "I have found that scientists' views of all of nature are highly influenced by their views of the small pieces of nature about which they are expert. Thus, for example, those steeped in decomposition [as in mathematics and physical science] tend to apply reductionism to all problems. Analytical modelers try to describe everything in terms of equations and computer programs."[87]

It is exceedingly strange that scientists tell us they have the key to reality or realities because of their rigorous and "objective" method. When we scan their pronouncements, we find them saturated with a subjectivity that most fundamentalists might well eschew as too emotional. J. V. McConnell, editor of *Skeptical Inquirer,* put it bluntly in a blistering commentary: "Scientists are just about as pseudoscientific when it comes to their own behavior (in and outside of the laboratory) as the crackpots are. For example, more than 30 years ago, Robert Rosenthal showed that scientists tend rather systematically to bias their laboratory observations. CSICOP [Committee for the Scientific Investigation of Claims of the Paranormal] often takes pis [related to ESP] researchers to task for not providing safeguards against 'experimental bias,' but I'd guess that less than 10 percent of the published studies *in all fields of science* include the use of similar precautions! I've seen more examples of data-fudging in the lab (most of it unconscious) than I care to remember—and some of the worst instances were by a Nobel Prize laureate in chemistry. Scientists are almost as likely to impose their values and expectations on 'objective evidence' as the crackpots are."[88] So much for objectivity.

We can trace the departure from any kind of objective view beginning with the famous "paradigm" that Thomas Kuhn, University of Chicago philosopher and author of the famous book *The Structure of Scientific Revolutions*, articulated more than a generation ago: "One of the things a scientific community acquires with a paradigm is a criterion for choosing problems that, while the paradigm is taken for granted, can be assumed to have solutions. To a great extent these are *the only problems* that the community will admit as scientific or encourage its members to undertake."[89] (Emphasis added.) How then can spokespeople for science maintain their belief in the objective nature of the scientific enterprise? Scientists seem to have as much flexibility in presenting their ideas as do fundamentalist preachers with their strange interpretations of the Bible.

Assuming that the unconscious choices made within the paradigm to which we give allegiance were objective in the sense that they have always been familiar to us, we still have practical considerations about the subjectivity of science. Richard Milton, a science journalist whose book *Shattering the Myths of Darwinism* is generally thought to have begun the intense modern debate about Darwin, observed, "All scientists make experimental errors that they have to correct. They naturally prefer to correct them in the direction of the currently *accepted* values, thus giving an unconscious trend to measured values. This group thinking has even been given a name: 'intellectual phase-locking.'"[90] This tendency is not a recent problem—it is a procedure so common as to be unquestioned. It most probably illustrates the intense desire of individual scientists to be accepted by their peer group and has nothing whatsoever to do with objectivity.

Radiocarbon readings are notoriously subjective, and they hide a bevy of transgressions. J. Ogden, the director of a radiocarbon dating laboratory at Wesleyan University in Ohio, pointed out that the scientist bringing samples to the lab is usually asked what the acceptable dates would be for the sample. When a date is obtained near the scientist's desired date, it is reported along with some variance to fudge the figures a bit. "It may come as a shock to some," Ogden said, "but fewer than 50% of the radiocarbon dates from geological and archaeological samples in northeastern North America have been adopted as 'acceptable' by investigators."[91] University of Arizona paleontologist Paul Martin implied that, in formulating his big game thesis (the idea that primitive hunters eliminated more than thirty species of megafauna), he had discarded samples with dates that conflicted with his preferred dates.

Read Stephen Jay Gould's *The Mismeasure of Man* for an examination of the high-profile scientists who altered evidence in studies of human anatomy to make their reports concur with then-accepted racial theories. Their intent was to demonstrate that the white race was the smartest and most important product of human evolution. Believing that a large brain indicated greater intelligence, scientists from Samuel Morton in the 1850s to contemporary Smithsonian scholars have been measuring human skulls in a futile effort to determine race. This belief is somewhat akin to Nazi beliefs about a master race, although today clothed nicely in big scientific words. Outside of showing outright fraud in a couple of instances, Gould generally adopts the posture that they were "men of their time." But their ideas were accepted and passed off as orthodox science. Have things changed so much that we can turn our beliefs over to scientific pronouncements? How much validity can there be in a scientific process that has these kinds of skeletons in its closet?

Finally we have Ian Barbour's frank admission that "personal motives, such as professional recognition and the securing of research grants, will tend to favor working within the paradigm. Institutions and individuals may have a greater stake in one theory than another. Rapid acceptance of a particular theory and resistance to a rival one may have complex, social, political and economic causes."[92] Science in America, after all, is as much big business as it is the objective independent inquiry into truth. Can we then take Paul Davies seriously when he says, "Science demands rigorous standards of procedure and discussion that set reason over irrational belief."[93] Or did Philip Slater, social critic and sociologist, have the right understanding when he noted that "it isn't Nature that evolves slowly and peacefully but science itself: The theory of uniformity is a projection of academia unto nature."[94] Davies is touting the party line; Slater is serious.

Establishment science today has a dominating presence and a wonderful public relations image. Consequently, everyone strives to be included under the umbrella of scientific respectability by finding a way to pretend that they operate on scientific principles. We sometimes have the appropriation of the name—as in "Recreation Science" or "Military Science and Tactics." We have been trained to believe that everything we do must have scientific endorsement, and once the label has been accepted, the activity, no matter how bizarre, has standing in our society. If we are looking for certainty—which should be the goal of any knowledge-seeking society— we have no more assurance of truth or objectivity from the science of today than we do from the fundamentalist preacher, or from the alchemists and witch doctors of past ages. How do we answer Kenneth Ring, a pioneer in the investigation of near-death experiences, when he says that there is "a kind of double standard by which the writings of physicists dealing with such speculative possibilities as quarks and tachyons are treated with respect, where the writing of parapsychologists on out-of-body experiences or reincarnation phenomena go unread or are simply dismissed."[95] Well, if we are honest, we simply say, "Our paradigm does not allow such things to exist. Perhaps someday, when we are less materialistic, we will be allowed to examine those things."

Whenever popular writers want to demonstrate how precise, logical, and rational science is, they always use physics as their primary example. Physics invokes images of brilliant mathematicians and thinkers investing a tremendous number of hours with complicated instruments that can measure exceedingly minute and remote things and arriving at precise formulas that describe physical realities. We remember the complicated mathematical formulas that describe the experiment and the findings. Big Bangs, clashing galaxies, minuscule particles, and exploding atoms are all images that characterize our view of physics. As we move away from physics and the hard sciences toward the social sciences, the less we hear about precision and logic. Indeed, we get fairy tales and hopeful scenarios. Many of the evolutionary explanations are reminiscent of Rudyard Kipling's *Just So Stories,* wherein frothy and delightful but wholly impossible tales are spun to explain why tigers have stripes or turtles have shells. And we tend to credit all the rigor and discipline of physics to other sciences that lack any semblance of objectivity

But even physicists do not operate with the objectivity that we are told exists in science. Scattered comments about the creative process in physics tell a much different story. "Scientists then often speak of the 'scales falling from the eyes' or of the 'lightning flash' that 'illuminates' a previously obscure puzzle, enabling its components to be seen in a new way that for the first time permits its solution," Kuhn wrote in *The Structure of Scientific Revolutions.* And he observed, "On other occasions, the relevant illumination comes in sleep. No ordinary sense of the term 'interpretation' fits these flashes of intuition through which a new paradigm is born. *Though such intuitions depend upon the experience, both anomalous and congruent, gained with the old paradigm, they are not logically or piecemeal linked to particular items of that experience as an interpretation would be.*"[96] (Emphasis added.) This description does not match the image we have of the hardworking, hard-thinking, rational scholar who deduces truth while eschewing anything smacking of mystical insight. And it is not what Paul Davies has claimed on behalf of science.

Unfortunately, Kuhn qualified his remarks later in his book, taking the side of the scientific establishment and negating his earlier insight. It is pitiful to read his confession: "Some readers have felt that I was trying to make science rest on unanalyzable individual intuitions rather than on logic and law. ... If I am talking at all about intuitions, *they are not individual.* Rather they are the tested and shared possessions of the members of a successful group, and *the novice acquires them through training as a part of his preparation for group-membership.*"[97] (Emphasis added.) Can this process be possible? Can we train people in our labs and universities to be intuitive as part of membership in a group? Can you test and share intuitions? And do *groups* have intuition? Was Kuhn borrowing from Carl Jung's theory of the collective unconscious? Of course not. We must recognize that Kuhn had to maintain his status within the scientific community. Admitting that intuitions play a role in the scientific endeavor would have introduced "mysticism" into the equation, and that would have been heresy. To leave Kuhn's thinking with a better taste in our mouths, let us remember his insight that "almost always, the men who achieve these fundamental inventions of a new paradigm have been either very young or very new to the field whose paradigm they change."[98] Minimally, they have been people on the fringe of the group.

Paul Feyerabend states the case more specifically: "Science is advanced by outsiders, or by scientists with an unusual background. Einstein, Bohr, Bonn were dilettantes, and said so on numerous occasions."[99] Pray tell, what institutional experiences do these thinkers share? David Foster, English philosopher, in his brilliant book *The Philosophical Scientists*, offers us a list of major accomplishments in physics:

Planck and quantum theory

Einstein and relativity theory

De Broglie and matter-wave equivalence

Schrödinger and wave mechanics

Heisenberg and the uncertainty principle

Pauli and the exclusion principle

He observes: "Not one of these was reasonable or common-sense, but all were true and *they worked*. They were the result of 'inspiration.'"[100] While the discoveries of modern physics, the most successful of our physical sciences, may be described using complex geometries and mathematics, the *source* of many important and fundamental theories in physics is mysticism.

Edward T. Hall, a well-regarded anthropologist, reported that Albert Einstein "did not think in words, nor did his important insights come to him in mathematical terms. Instead, he had physical images coupled with visual images that represented complete entities (systems) which then had to be laboriously broken down and translated into mathematics and words."[101] Needless to say, Einstein had an intuitive mystical insight and then expressed it in a mathematical form that his colleagues could understand. It did not happen the other way around, despite Kuhn's belief.

According to David Foster, "the great breakthroughs in science were almost all of a mystical nature in which an emotion of 'problem solved' preceded the solution in Queen's English or mathematical symbols."[102] We can trace even the Cartesian mind/matter separation to a mystical experience. On St. Martin's Day in 1619, René Descartes had a vision of a mechanical world that became the basis for his philosophy. He believed it had been "channeled" (as we would say today) by the Angel of Truth, and he undertook a pilgrimage to the Shrine of the Virgin at Loretto two years later as a means of giving thanks for the insight.[103] Discussing the appearance of intuitive information that inspired the creative work of Wagner and Coleridge, F. David Peat, English quantum physicist and psychologist, noted that Werner Heisenberg, author of the indeterminacy theory in physics, perceived the basic insights of quantum mechanics after walking and swimming while on vacation from the intellectual controversies of the University of Göttingen.[104] I stress the importance of mysticism strongly because our society, with our entrenched belief in scientific folklore and its myth of rigorous rationalism, generally discounts the fact that mystical insight has played a critically important role in the creation of new scientific theories.

If we compare religious experience with scientific thinking, we arrive at the realization that the processes are much the same; we should not laud the "objectivity" of the scientists while deriding the "subjectivity" of the religious thinkers. Physicists and many other scientists receive their insights in pretty much the same way that mystics and tribal medicine men do. The only real difference is that physicists use a different descriptive language—mathematics—and seek to articulate their experiences in ways that can be verified by others prepared to follow their formulations. Mystics describe their experiences in parables and stories, in analogies and symbols, while medicine men express their insights by performing ceremonies. In religion, as in science, other people can follow the lead of religious personalities, perform ceremonies and rituals, and have religious experiences in a similar way.

Heisenberg points out the nature of the difference between the two ways of gaining knowledge, admitting subjectivity but using much more familiar terms: "The most important difference between modern science and ancient natural philosophy lies in the method employed. Whereas in ancient philosophy the empirical knowledge of natural phenomena was reckoned sufficient for drawing conclusions about the underlying principles, it is a characteristic feature of modern science to institute experiments, i.e. to put specific questions to nature, whose answers are then to yield information about its laws."[105] Since we always trace concepts back to those ancient philosophers when we articulate them, we should admit that they were as intelligent as we are but used a different approach to the problem. Indeed, they may have been smarter than we because they lacked the instruments and mathematics that we use and yet arrived at similar conclusions.

We seldom see the basic method of gaining knowledge stressed in the articulations of scientific doctrine today. When we have to explain ourselves in nonmathematical language, we are restricted to images that are familiar to our everyday life—"waves" and "holes" and "strings"—so that the results are not radically different from those reached earlier or by other people. We usually talk about the *behavior* of entities, and the process of observation is simply that of looking at nature and describing what we see in an efficient manner. Scientific experiments simply narrow the focus of our inquiry and force nature, in a Baconian fashion, to show us how it behaves under very carefully defined circumstances. We are not using a more sophisticated procedure to secure greater truths than would ordinarily be available to us. And our results are actually restricted in scope and application.[106]

Physicists recognize that this method is critical, but in their efforts to describe in simple language what they have concluded, they often use ancient philosophy as their framework of interpretation. Thus Werner Heisenberg observed: "The elementary particles of modern physics can be transformed into each other exactly as in the philosophy of Plato. They do not themselves consist of matter, but they are the only possible forms of matter. Energy becomes matter by taking on the form of an elementary particle, by *manifesting* itself in this form."[107] "Manifest" is an excellent choice of words here because it describes what is clearly apparent on each side of the experimental equation. We begin tracking bits of matter and conclude there is no matter, only a very complex field of energy that represents it. We will return to this concept later and discuss its possibilities as a properly descriptive term for the origins of biotic systems following a planetary catastrophe.

Heisenberg also observed that "it is no longer the objective events but rather the probabilities for the occurrence of certain events that can be stated in mathematical formula. It is no longer the actual happening itself but rather the possibility of its happening—the potentia, to employ a concept from Aristotle's philosophy—that is subject to strict natural laws."[108] But what is potentia? Here is something not material, existing only in the realm of logical possibility, about which nothing can be said except in an ex post facto manner. We have lost the possibility of prediction with accuracy, a goal cherished by centuries of scientists. If situations contain all manner of potentia with respect to their physical constituents, and if nothing can be predicted with accuracy prior to the event, are we not in the realm of mind? Are we not seeing a part of a creative process in the moment prior to the realization of the event? Would not the interpretation of this condition be a wholly subjective process for the scientist and become primarily objective only when measuring the *results* of the experiment? This situation may be the frontier of understanding that we cannot cross.

So what is it that "objective" science really knows? In the public arena we are assured beyond doubt that the results of our experiments give us truths denied to earlier and less worthy generations. Once we remove the image of science as an absolute source of truth (against the objections of Barbour and Davies), we find reasonable statements about the *limitations* of our knowledge, not the triumphant shouts of modern alchemists. We began with the conviction that nature "exhibits herself more clearly under the trials and vexations of art [mechanical devices] than when left to herself"[109]—an idea first promulgated by Sir Francis Bacon to justify the new experimental method—and eventually arrived at the ambiguity noted by

Percy Bridgman, who expressed surprise and consternation when modern physicists discovered that their concepts merged together at the subatomic level, leaving them an increasingly sparse vocabulary.[110] We may begin by trying to force some data from nature, but when we come to understand what we have done, we find ourselves looking at our own minds.

Thus it was that physicists began to move away from the idea of possessing absolute knowledge to a more realistic interpretation of their activities. "When we speak of the picture of nature in the exact science of our age," Heisenberg said, "we do not mean a picture of nature so much as a picture of our relationships with nature."[111] And he cautioned: "We have to remember that what we observe is not nature in itself but nature exposed to our method of questioning."[112] This limitation clearly leaves room for other expressions of nature—and of our relationship with nature—that cannot be described mathematically but nevertheless should have equal standing with scientific findings. Here Kenneth Ring's complaint about scientists who deal wholly with material things, but who deride scientific work in the area of intangible things, is valid. It is only because we still believe in a material basis for the cosmos that we deny the reality of psychic, nonmaterial events.

If we take one of the most basic concepts of modern physics, the quantum wave, and examine it, we find uncertainty: while the measurable results are precise, our ability to describe what happened is not. Our language seems unable to transcend the old Newtonian framework, and we must use familiar images to convey our understanding. Fred Alan Wolf, American theoretical physicist, pointed out that the quantum wave itself has never been seen. It is merely a hypothetical solution to a mathematical problem. In other words, "Whether they are real and objective or not, all physicists must think of quantum waves 'as if' they were out there in space and moving in time like any common waves you might witness, in order to give meaning to their computations of the correct mathematical probabilities of specific events in space and time."[113] We must admit that at the macro and micro levels of investigation, space and time are merely concepts. At the meso level they are real because they enable the physical to become manifest.

The same can be said about the popular "black holes." Heinz Pagels, philosopher of chaos theory, suggested, "They are now routinely invoked by theorists to explain almost any new observation in astronomy that requires a huge energy source in a small region of space."[114] And Harold Booher, guest scientist for the Army Research Center, offered a more pointed critique:

In order to make the Big Bang possible within the parameters required for the open expanding version, the Universe must now be defined as comprising *99 percent dark matter*. It certainly is not obvious why the visible universe filled with stars, galaxies, and glowing gases must be relegated to a nearly negligible portion of the matter making up the "real universe." Big Bang faith requires unobservable matter made up of invisible particles because otherwise the Universe could not have formed in accordance with the Big Bang model.[115]

It must be terribly disconcerting, then, for some proponents of science to be forced to admit that much of what passes for scientific certainty is simply the personal belief that entities exist because they help explain mathematical equations. As there are changes in mathematics, particularly geometries and differential equations, so also do the entities that we are told compose our physical universe change. Remember the famous ether? Where has it gone? It was killed by relativity! But it may someday return— if it proves useful to describe in plain language what we are able to demonstrate mathematically.

In April 2002 Paul Steinhardt of Princeton University and Neil Turok of Cambridge University proposed a theory challenging the Big Bang's ability to account for the fact that the expansion of the universe appears to be speeding up. According to Charles W. Petit, writing in *U.S. News and World Report*, the new theory relies on string theory and M-theory and suggests that the universe expands and thins, a force field collects and contracts it again, and it recreates itself. So we often have a choice of scientific theories, and relying on the pronouncements of experts is not the best way to do business.[116]

Our best thinkers, however, continue to suffer from beloved Harvard philosopher and mathematician Alfred North Whitehead's "misplaced concreteness"—the idea that we take familiar things as real even though we know that they are not. Paul Davies concedes that "what we today call the laws of physics are only a tentative approximation to a unique set of 'true' laws, but *the belief is that as science progresses so these approximations get better and better with the expectation that some day we will have the 'correct' set of laws*. When this happens, theoretical physics will be complete."[117] (Emphasis added.) By every reasonable, logical definition this goal cannot be reached. If what we know is our relationship to nature, as long as there are humans and mathematics, we will be struggling to express our experiences whether in personal emotions or laboratory results. There can be no "correct laws" because everything is a statistical expression, and ways of expressing and describing statistical happenings can change.

David Lindley, in his book *The End of Physics*, modifies Davies's optimism: "Modern cosmological theories are built on ideas that have no proven validity, if one insists on the old-fashioned standard of empirical science. The hope of the cosmologists is that, in the fullness of time, observations and theory will come together in one particularly neat arrangement so elegant that it will be persuasive *despite the lack of solid evidence*."[118] (Emphasis added.) But the basis for science at many levels is the description of the empirical world supported by considerable evidence. Indeed, when we leave empiricism we must of necessity enter the field of theory, and the difference between theory and doctrine is not always made clear. Although this is a hopeful attitude, it is not radically different from religious thinkers espousing beliefs because they are phrased with great elegance. Surely Tertullian, an early Christian theologian, has some profound influence on our modern thinking in this respect. "I believe because it is impossible!" he said.

Even accepting the fact that we cannot know ultimate realities, we do have theories as fruitful paths that can lead us to new insights about the universe and ourselves. One path that would be productive would be reconciling what we think we know about the universe—the empirical data and interpretations of subjects we can describe mathematically—with our aesthetic, emotional, and mental apprehensions of the same universe. The ultimate goal would be to reconcile science and nonscientific human experience. Primary in this larger arena would be the experiences we call religious. It would involve revising our interpretation of data from the social sciences that has been gathered under the premise that human behavior, values, and beliefs can be treated objectively. We should ground our knowledge in a larger epistemological scenario that would enable us to admit the mysticism of physicists as well as the conclusions of nonscientific mystics.

"Sounds, smells, colors, and feelings are nowhere to be found in mathematical physics," Rupert Sheldrake, British advocate of the "morphic resonance" theory, reminds us, "because physics ignores everything that cannot be quantified."[119] At the present time we pretend that the data that is subsumed and explained under the rubric of science is absolutely correct in both its gathering and its interpretation. Some scientists believe that every phenomenon can be reduced to its origins in matter: that near-death experiences, for example, are nothing but aberrations in the chemicals in the brain, or that generosity is a virtue bestowed by genetic material. Apologists for these extreme viewpoints contend they are using the "scientific method," although the basis upon which they form their hypotheses is little more than the application of nineteenth-century materialism.

There is no need to abandon science because we have acknowledged that it is simply our best guess, though a mighty sophisticated one, as to what exactly we are facing when we try to make sense of the physical world. Indeed, American society is so constituted that it would be impossible to abandon the scientific enterprise. In the eyes of most Americans, it is considered outrageous even to criticize science. We can, however, modify science's claim, voiced so eloquently by Gould, that it alone possesses absolute and truthful knowledge.

There is no need, either, to modify religious thinking about the world to conform to the latest scientific findings. Especially, we should not tie our religious insights to a worldview that is in constant change and reformulation. Religious thinkers should continue their search for an adequate expression of ultimate reality just as scientists are doing—"ultimate" here being what we can with confidence express in meaningful ways. Religious thinking would do well, however, to deal with the religious experiences of people today rather than continue to probe the doctrinal expressions of the past. Above all, we should balance the scales in the courtroom so that representatives of science do not pronounce as "fact" what is tentative and tenuous.

CHAPTER FOUR

THE LOGIC OF EVOLUTION

WE HAVE SEEN THAT EVOLUTION is not a useful concept in micro-biology except as a password that entitles the scientist to publish studies. It is, however, vitally essential in paleontology since that discipline involves the interpretation of fossil data. Without the framework of endless time and the unproven belief that minor changes in body size demonstrate evolution, paleontology would have very little to tell us about Earth history or even about past organic life. We are so accustomed to hearing that evolution can be demonstrated in the fossil record that we approach the analysis of the fossil record with eyes expecting to find evolutionary evidence. This attitude is not science; it is belief.

The most telling critiques of the concept of evolution have not come from the creationists, however, but from two lawyers: Norman MacBeth and Phillip Johnson. MacBeth began reading evolutionary texts some years ago when he was restricted from practice by a severe heart condition.

Accustomed to dealing with formal rules of evidence, MacBeth found that evolutionary thinking was filled with unwarranted assumptions and conclusions. His book *Darwin Retried* is a masterpiece of analysis and shows that evolution is a descriptive term for very confused efforts to force the interpretation of fossil evidence into a predetermined mold. More recently, Phillip Johnson, a law professor, in his book *Darwin on Trial*, discussed additional shortcomings of evolutionary doctrines. Although he has recently been accused of being a creationist, his book is clearly the work of a disciplined mind simply asking for a clear explanation of why evolution is so popular when it is so illogical.

Three other contemporary writers also weigh in against evolution and offer devastating critiques, and it may have been their books that encouraged some teachers to consider intelligent design as a possible alternative to evolution, thus stirring up resistance to the subject. *The Facts of Life: Shattering the Myths of Darwinism* by Richard Milton offers some important insights into evolutionary thinking and shows that in many instances, it is simply an explanation without explanatory powers, a mantra to be chanted rather than a scientific doctrine to be used profitably. Harold Booher's *Origins, Icons, and Illusions* covers an even larger circle of concern and prepares the way for scientists to move beyond the present impasse toward a greatly enlarged view of biotic existence. Jonathan Wells's *Icons of Evolution* illustrates how textbooks are misrepresenting the subject by repeating old shibboleths and failing to correct images of the stages of development in illustrations of the evolutionary process. It seems ludicrous to give states low marks in their treatment of evolution when the textbooks themselves are in error.

The criticisms of these thinkers, taken together, offer irresistible arguments and considerable evidence to support the idea that we need a new paradigm with which to deal honestly with the fossil data. Their critiques are important because they require of evolutionary explanations the same rigor required of other sciences. If one were to transfer evolutionary logic, particularly that of the theory of punctuated evolution, to other sciences, we would have nothing substantial to build on. When we read evolutionary apologists, then, we should view their offerings critically, with the same demands we place on other sciences. One of the tenets of the scientific method is that concepts, doctrines, and dogmas should have some kind of predictive power. Many thinkers agree on a basic criticism of evolution: it is always applied *after the fact* to work data into a predetermined storyline. Milton says: "As a theory, natural selection makes no unique predictions but instead is used retrospectively to explain every outcome; and a theory that explains everything in this way explains nothing."[120] We can examine how evolution explains nothing by looking at purported explanations.

Donald E. Tyler, science writer, in criticizing evolutionary explanations, observes, "If an animal is dull in color, it is claimed that it survived because of camouflage. On the other hand, if the animal is brightly colored, its survival is attributed to increased sexual attractiveness or that the color served as a warning to enemies."[121] How do evolutionists choose which of these competing explanations represents evolution? There is no good answer except the personal preference of the scientist. George Gaylord Simpson, for half a century the reigning American evolutionist at Harvard, writing in 1953, explained that "development of armor, is, for instance, a frequent adaptation for defense, but it lessens the possibility of rapid motion, which is also a frequent adaptation for defense in other animals and has generally a wider usefulness than has armor."[122] So why do some animals have armor and move slowly while others lack armor and move quickly? We don't know the answer, and Simpson can't tell us.

The problem is that we can always invent a plausible adaptive advantage for an observed or supposed trait. Fine—but is this science or folklore? Simpson admitted in an offhand manner, "It is not surprising that opinion as to the adaptive significance of differences between taxonomic groups varies from belief that such differences have no adaptive value to the claim that they are always adaptive."[123] Is there a right or wrong answer here, or only opinions? Is this thinking not akin to the opinion of a cardinal as opposed to a parish priest? If we look carefully at the workings of evolutionists, we will discover that they are engaged in a massive task of classification. The so-called evolution of species is primarily the transfer of identifiable species from one category to another and the creation of new categories to explain morphological resemblances. Again, George Gaylord Simpson can be summoned to explain the process of classification: "Our recognition of a higher category is *ex post facto*, as is our designation and placing of it in the hierarchy. The Cricetidae (a family of mouselike rodents) are a family because they have become so extremely varied. If there were only a few genera or species of Cricetidae they would be members of the Muridae."[124] So it may not be an evolutionary family tree at all that we are viewing, it may simply be one species with a great many variations or another species with few variations.

In fact, the much-celebrated charts we see illustrating the descent of all living beings from the single cell are mere fictions, and Jonathan Wells asks why they are included in textbooks when scientists themselves know better. Simpson admitted, "Diagnosis of a higher category is not a description of any ancestor or of a type of organization ever embodied in or definitive of an actual organism or population. It is an abstraction of the broader adaptive features present, usually with variations, in the members of a category."[125] So family trees exist primarily in the minds of the evolutionists.

The problem is worse than we might suspect. Simpson says that "the assumption that genera, say, cover more or less equivalent amounts of total evolutionary change seems sufficiently valid *when the genera are closely related and have been defined by a single skillful student or by more or less equally able students using similar criteria.* That assumption becomes, however, increasingly uncertain when the groups compared are more distantly related and more dissimilar and *when their definition has been done by students with different taxonomic tendencies and using different criteria.*"[126] (Emphasis added.) Here we do not see nature telling us its secrets but rather skillful students creating interpretations of the data. Simpson's confession comes from his early writings in the 1950s, but we have no reason to doubt that the *method* of explanation has been refined in the decades since then.

Indeed, Robert Bakker identifies the same sins when paleontology adheres to traditional methods of interpretation. Surveying his own discipline, Bakker confessed, "Paleontologists had somehow arrived at a view of evolution which I call the hub-and-spoke syndrome. Each major evolutionary line supposedly originated in one primitive, unspecialized stock, an evolutionary hub. Subsequently, all the advanced lines evolved outward—like separate spokes of a wheel—to form that hub. Under the influence of this conception, paleontologists tended to invent wholly imaginary groups to serve as the ancestors for their grand theories. Poorly known fossils, represented by fragmentary skeletons, were often elevated to the status of 'common ancestral stocks.'"[127] How, then, can evolution be regarded as a *fact*? Is it not really a belief comparable in every way to a religious belief? Does this method of "studying" evolution prevail today? You bet it does, except that the evolutionary "bush" has replaced the "hub-and-spoke." Evolutionary classification, then, is not exact, nor is it a science; it exists primarily in the minds of its practitioners. When two authorities clash, the senior and more prestigious scholar's views are considered orthodox until the younger scholar gathers more data or the senior scholar dies. Was there misrepresentation in those Arkansas and Louisiana courtrooms when scholars assured the judges that evolution had been proven beyond doubt?

We are always comforted by the thought that at the species level we do not have to deal with speculations about families and family trees. Species, we are often told, are defined by their ability or inability to interbreed, although a recent summary of the current definition of species seems to refute the breeding requirement.[128] The limitation of breeding capability might hold true for some but not all living species, but it is difficult to understand how one could know if a fossil species could or could not breed with another similar species. We cannot recover DNA or chromosome numbers from stone fossils, so whatever criteria we might use to define different species would have to revolve around body shape or our projections of what organic parts are similar in shape or possible function with each other. Projections based on these elements are hardly reliable since, again, they are based on our definitions and not on anything in nature itself.

Evolutionists are so confident in themselves that they often confess the most dreadful sins without understanding that these sins weaken their case dramatically. Stephen Jay Gould writes that "taxonomists tend to fall into two camps—'lumpers' who concentrate on similarities and amalgamate groups with small differences into single species, and 'splitters' who focus on minute distinctions and establish species on the smallest peculiarities of design."[129] Of course this difference in perspectives leads to some embarrassing statements that evolutionists do not parade in front of us when their methods are called into question. They do not admit such shenanigans in courtrooms either. Gould says, "Agassiz [the first great American advocate of evolution] was a splitter among splitters. He once named three genera of fossil fishes from isolated teeth that a later paleontologist found in the variable dentition of a single individual."[130]

There is a retrospective correcting process in evolutionary writing in that many contemporary spokespeople cite the failures of their predecessors in explaining their own conclusions. They fail to realize that revealing these shortcomings calls evolution into question and shows how simplistic and absurd the evolutionary explanations are. Matthew Bille, publisher of the *Exotic Zoology* newsletter, in his book *Rumors of Existence,* noted that there were once eighty-six species of brown bear while the splitters were prominent and that later zoologists, who were lumpers, reduced that number to a single species with four subspecies.[131] One can only wonder what loyal evolutionists thought when their brown bear family bush was denuded of its branches. Even worse, Gordon Rattray Taylor, in *The Great Evolution Mystery,* cites an example of the complete nonsense of evolutionary exposition. "The French taxonomist Locard classified the freshwater mussels of France into no fewer than 251 species on the basis of their shell forms and colours," Taylor reported. "Today all 251 are regarded as a single species. Another man recognised 200 species of snail in Hawaii in 1905, whereas a three-man team who went there seven years later put the number at forty-three. Darwin's finches themselves have been the subjects of this treatment, having been classified into more than thirty species by a taxonomist who commented later that he might as well have called them all one species."[132] If this kind of activity characterizes how evolutionary theory demonstrates gradual change, how on Earth can we get the evolutionists to reach some kind of compromise on whether the data should be split or lumped so that some stable count of species can be achieved?

Part of the technical vocabulary, and certainly an integral part of evolutionary interpretation of fossils, is to identify a "trend." "Trend" means some physical characteristic that is unique to a species and is interpreted to indicate evolutionary development according to orthodox Darwinian theory. Again we face the nebulous explanations that deprive us of understanding. Simpson said, "It seems probable that any new evolutionary trend might have been initiated on the basis of existing variability, or again that it might not in a given case."[133] But Simpson failed to realize that this arbitrary act of identifying trends does not explain evolution. We want to know *when* you apply the concept of trends and when you don't. As he would have it, identifying trends is a whimsical choice that depends more on the imagination of the scientist than on the data in the fossil record.

When Simpson applies the concept of "trends" to his data, the result can become very mysterious indeed. He writes: "Often we do not know enough about the functional aspects of a trend to know why it ceased to be advantageous, but in many examples this is quite clear. For instance, under normal circumstances a wild horse's grinding teeth last until the animal is senile from general organic aging and past breeding age. There would be no advantage to the species in greater hypsodonty so that senile horses died with unused grinding capacity."[134] Is this science or merely nonsense? Simpson apparently isolated the grinding teeth from the rest of the animal and at first believed that good teeth were advantageous for survival and therefore represented a "trend." He then realized that old horses with younger teeth are an absurdity, that the condition would not be advantageous, and then cited his own confusion as an illustration of the idea of evolutionary trends. Common sense dictates that as an animal grows old, all of its organs and bones age at about the same rate. He further confuses us, and himself, by including the phrase "under normal circumstances" as if there were sometimes abnormal circumstances in which teeth did not age as fast as the rest of the animal.

Another of his startling evolutionary insights was that "some truly genetic character correlations may be rather obvious and indicate plainly only that the characters, as analyzed, are parts or aspects of a larger character. Thus if an organ becomes larger as a whole, so will its height, width, and breadth in close correlation with each other."[135] The observation that as organs grow larger, they do so proportionately should not be seen as a magnificent accomplishment of evolutionists. This phenomenon can be observed by almost anyone watching something grow. Simpson was regarded as one of the giants of the twentieth century in evolutionary thought. It is somewhat discomforting to realize that several generations of biologists read his books, came across insights like horse's teeth, and uttered not a single complaint. Did fundamentalist students reading the Bible ever have fewer questions?

When we boil down the rhetoric of evolutionists, we find a frank admission by Steven Stanley: "We do not know exactly what evolutionary changes *other than increase in body size* are most likely to take place throughout a populous species. Presumably such changes should be ones which are likely to be of value to all populations, regardless of peculiarities of the environment."[136] (Emphasis added.) Now, doesn't this statement mean that apart from growth of body size, which should be anticipated, we really know very little about how organisms originate or develop? While we hear glowing tributes to evolution from many sources, the fact that we are dealing only with slight variations in size and color in living organisms should cause us to reduce our praise a decibel or so.

Robert Wesson brings up an interesting observation in his book *Beyond Natural Selection:* "If the same criteria were applied to humans as naturalists apply to animals, a dozen persons might well be placed in a dozen different species on grounds of build from pygmies to lanky Nilotic herdsmen and squat Eskimos, distribution of fat (buttocks, bellies, breasts), hair color (black to red to flaxen), hair shape (from tightly kinky to straight), different body hair, skin color, eye color, lips, noses, and so forth—traits for the most apart seemingly without adaptive significance."[137] Some anthropologists, however, continue to classify human races and ethnic groups based on the measurements of their skulls even though such a practice was discredited more than a century ago. It was testimony on skulls that six scientists offered in evidence in the famous Kennewick Man case involving the identification of the possible race of a 9,400-year-old skeleton found in the Columbia River.[138] The argument was that a narrow skull meant he was a white man. Without DNA and chromosome counts, when we have just the fossilized bones, aren't we simply imagining evolutionary change? Could not the different "species" we identify using bones alone simply be variants of a basic animal?

Here lies the problem with fossil interpretation. We really have no idea what the animals looked like in real life—absent, of course, the dramatic reconstructions done by moviemakers. Our identification of evolutionary "trends" must rest wholly on changes in body size and bone shape. Niles Eldredge points out the great difficulty we have in identifying any kind of speciation. We know of course that both lions and tigers are members of the "cat" family because we have identified them as such. But "lions are organized in prides (with some loner males), while tigers, usually living in denser vegetation than lions, go it pretty much alone (males—and females once the cubs are old enough). *Yet without their skins, by bones alone, competent anatomists tell lions from tigers, if at all, only with great difficulty.*"[139] (Emphasis added). How often does this mistake in identity occur in our articulation of evolution when we are dealing with fossils allegedly millions of years old? With some very rare exceptions, we do not know fossil skins or colors; we know only the similarity of bone structure. Our talk about fossil "species" is therefore little more than pure speculation.

How did evolutionary biology get into such ludicrous shape? Our accusation earlier—that evolution tries to answer a basic religious question—now makes more sense. We have tried to substitute a mindless, directionless process for divine creative wisdom. And we have been led to draw meaningless conclusions. At least for a spell, we had 86 species of brown bear, 251 species of mussels, and 200 species of snails while the splitters occupied chairs of great prestige in the universities. Then, as scholarly fashions changed, we lost all but one species of bear, 250 species of mussels, and 157 species of snails by the simple matter of adopting more accurate definitions. That is a loss of 492 species that are now not part of the facts of evolution. We have had a 91.6 percent species loss simply because of more careful classification. How could you ever determine an "endangered species" with a track record like that? Should we, perhaps, put the definition of species to a vote?

Before the adoption of evolution as the framework for interpreting the planet's past, we were limited to Noah's flood as the primary agent of extinction for fossils found in geological strata. Since evolution took immense amounts of time, the need for endless geologic time followed. Geology and evolutionary biology then began an incestuous relationship that remains the dominant paradigm even today. A substantial percentage of geological strata are dated by the index fossils found in sedimentary rocks. No thought is given to critiques of this tautological condition wherein strata are arranged according to the complexity of fossils and fossils are arranged according to supposed geological strata. Although this situation does not involve embarrassing situations for geology, it requires evolutionists to dance a merry jig in creating a framework for explaining fossils.

Supposedly all forms of life are subject to the great evolutionary process of growth and change. But the fossil record does not show this growth or change apart from the increase in body size. If we look at the statements of evolutionists regarding the fossil record, we discover some astounding things. Most important is that fossils show an amazing stability over very long periods of time; entirely modern species are classified as millions of years old. We also have what we call "living fossils"—creatures still alive today that are believed to have originated millions of years ago. Let us look at a number of statements from evolutionists about these phenomenal creatures who have defied the basic rule of evolutionary life.

Sir Fred Hoyle and N. Chandra Wickramasinghe, in *Evolution from Space*, observe that "the forms of butterflies and dragon flies were set more than 100 million years *before* the emergence of birds. Yet without change they survive the biological challenge and they do so even though from a purely visual point of view they are almost blind to the attacks of their predators."[140] It should be somewhat disturbing to evolutionists that butterflies as prey existed long before birds existed as predators. Was the world for many millions of years overflowing with butterflies until the birds evolved?

Steven Stanley rhetorically asks: "What has happened to the bowfishes during their long history of more than one hundred million years? Next to nothing! During the latter part of the Cretaceous, bowfishes became slightly elongate, but during the entire sixty-five million years of the Cenozoic Era, they evolved only in trivial ways. Two new species are recognized, but these differ from their late Cretaceous ancestors only in subtle features that represent no basic shift of adaptation. The bowfins of seventy or eighty million years ago must have lived very much like their lake-dwelling descendants do today."[141] Bowfishes apparently led the resistance movement against evolution for millions of years. Or we simply don't need those millions of years. It would be easier to simply note that bowfishes in the strata are like bowfishes today—period.

Many other species exhibit the same resistance to evolutionary change. Although this is tedious in its repetition, I cite more examples used by critics of evolution to illustrate how widespread the problem is. Robert Wesson gives a roll call of species that have changed very little over vast stretches of time. "Over a longer span, ants preserved in amber 25 million years old look very much like those of today; some species are difficult to distinguish from modern descendants. ... Lungfish go back 350 million years, and horseshoe crabs (*Limulus*) have changed little, at least in their skeleton, for that time or longer. Some brachiopods (*Lingula*) are apparently unchanged, at least in their shells, for 450 million years. Other holdouts of bygone ages include crocodiles, some turtles, and a variety of bony fish and sharks. A number of plants, such as ginkgoes, cycads, horsetails, and club mosses, are at least 100 million years old."[142] How does this interpretation differ from the creationists' belief that, once created, species do not change significantly over long periods of time? It differs only in the phantom millions of years that the evolutionists believe have passed.

Harold Booher cites a study of insect fossils in German, Russian, and Chinese paleontology conducted by Conrad Labandeira of the Smithsonian and J. John Sepkoski, Jr., of the University of Chicago. Of 1,293 fossil insect families, 84 percent of the insects living in prehistoric times are living today.[143] And Steven Stanley admits, "Every bivalve lineage that has previously been assigned to a single species was found to encompass virtually no net directional change over any time span considered (up to 17 million years). Furthermore, two lineages that previously had each been divided into ancestral and descendant chronospecies were found to encompass virtually no net directional change: they deserve placement in a single species."[144] What would a completely honest roll call of species produce? Would we not have to admit that we can find virtually no change over the hundreds of millions of years we believe organic life has been on Earth? This material was obviously unknown to the courts that had to decide between evolution and creation science. How can so many scientists support evolution knowing what the situation really is?

What is occurring here when we can make long lists of creatures that have remained basically the same over a very long period of time? The most commonsense approach would be to ask if the timescale isn't a bit off kilter. How do we *know* that these creatures are really as old as we believe? Well, we estimate the time it would take to create geological strata, or we date the strata using fossils and project millions of years during which the majority of fossils remained stable. But these are not "real" years—they are estimates. To prove the value of our estimates we devise all kinds of measuring devices and calibrate the results of those tests against our already accepted timescale. The whole process is a fascinating exercise in self-delusion, and if the creationists tried such a silly chain of reasoning we would greet them with energetic ridicule.

Niles Eldredge offered an example of the precise calculations involved in paleontology. Speaking of a quarry in Morrisville, New York, containing trilobites, he wrote: "Even if we imagine that 1 centimeter a year were deposited—a rather rapid rate of accumulation, but not an utterly unreasonable estimate—we are still talking about 1,000 years for 35 feet to accumulate. That would be the least amount of time for that much sediment to be deposited."[145] The mathematics of deposition sound impressive and scientific, but there is a massive gap in his reasoning. Eldredge does not allow for the possibility that all kinds of sediment might have been dumped on these unfortunate trilobites by a rigorous flash flood or that a series of flash floods might have taken thirty-five feet off the top of the formation with equal ease. Can we think of any place on Earth where sediments could have accumulated uniformly for a thousand years without some kind of disturbance that would negate our mathematics? Most of the calculations of geologic time do not admit that anything except gentle sedimentation occurred—and that over time periods so long that most locations would not be recognized in their original state. Surely ordinary changes in deposition would have occurred that would have invalidated any simple measurement or estimate of the passage of time.

What are some other ad hoc explanations used by the evolutionists? Vestigial organs come immediately to mind. We sometimes find strange appendages on animals that are interpreted as the remnants of former organs. We can try to link species to each other in a quasi-family tree based on the idea that the strange appendage represented organs that fell into disuse and were discarded. The sad fact is that vestigial organs are usually organs whose function is unknown. Harold Booher explains how evolutionists use the concept of vestigial organs: "When identified in the animal world solely to support evolution, they, like the whale rear legs or those of the python or boa, are often not vestigial at all. The spurs (vestigial legs?) of the serpent are useful now, both in killing its victims and in propelling it along the ground. With humans, organs classified as vestigial in the past were often done so in error simply because their functions were unknown at the time. Such important glands as the tonsils, thyroid, pineal, and pituitary have at one time or another been labeled vestigial."[146] Thus glands, once important concepts in explaining evolutionary developments, are no longer available to prove the steps of human evolution except by people still living in past theoretical fantasies.

Strangely enough, bats have proven a major problem for evolutionists. George Gaylord Simpson used them as a means of calculating the tempo of evolution. As described by Norman MacBeth: "Working from what he [Simpson] knew of the fossils and time sequences, he could see that the bat's wing, for instance, has changed very little since the middle Eocene (about 100 million years ago). If its earlier evolution had proceeded at the same slow rate, its total time of development would be greater than the age of the Earth, a manifest absurdity. Therefore Simpson concluded that in the early days the rate for bats must have been ten to fifteen times as fast as later."[147] How would the bat endure this rate of change, and do we have any transitional forms to support this faster rate of evolution? Philip Kitcher, in *Abusing Science,* posed a tough question: "How did bats continue to exist before they had wings? How did they manage while the wings were developing?" And Kitcher offered an orthodox answer to his questions: "The answer is obvious. The ancestors of the bats occupied a different place in the environment. Once the bats had evolved they were able to exploit environmental space that was inaccessible to their ancestors."[148]

Now, a bat without wings might well simply be a mouse. It's difficult to imagine the kind of environment that wingless bats could occupy without their simply being some other kind of rodent, nor can we determine the reason for their growing wings in order to occupy a new environment if the species was surviving in its original environment. Flying would give access to a new environmental space, and since flying must be considerably more efficient than running on the ground, we can only wonder why every creature did not sprout wings like the bat. Again, we have no empirical data, no good answers; we have only the authoritative speculations of a professor to verify this belief.

So we have not answered questions but merely substituted indecipherable gobbledygook for a real solution to our problems. We can only wonder what impels evolutionists to pose unanswerable questions like these and offer answers that cannot possibly explain the situation. Adrian Desmond articulates the reverse of Kitcher's bat questions in his book *Hot-Blooded Dinosaurs*. Desmond writes, "Birds, like mammals, had been a subjugated race during the Mesozoic and remained an inconspicuous element of the fauna. They became proficient flyers but were unable to exploit the ground as runners for fear of being hounded by the giant predators."[149] Now, if birds could fly, what on Earth would have possessed them to want to run along the ground? How could a large predator ever catch a small bird that could both run and fly? The teeth of tyrannosaurs were certainly not designed to consume birds, and the idea of a lumbering giant or even a much smaller dinosaur chasing birds boggles the imagination.

Why does Desmond insist on this scenario? He wants to connect dinosaur fossils with birds to sketch an evolutionary progression. Witness his view on this point: "Many birds took refuge in the sea and some, like Marsh's Cretaceous *Hesperornis,* reduced their wings to stubs and developed strong paddling legs. Only with the passing of the dinosaurs at the end of the Cretaceous could birds take to the empty plains. Since the fastest way to progress over flat terrain is to run, we find even early in the Tertiary large and long-legged ground birds."[150] Now, is there anything realistic about this interpretation of the fossil evidence? Birds exchanging wings for paddles and then coming out of the sea again, eschewing wings, and deciding to develop running abilities because running is faster than flying? Birds had to make a lot of evolutionary decisions in those days. Why didn't all other creatures, subject to the same predations by the dinosaurs, take similar precautions and change also? But even worse, why would birds ever decide to flee to the sea?

Do we even need the hundreds of millions of years for life to evolve? Steven Stanley tells us, "Within perhaps twelve million years, most of the living orders of mammals were in existence, all having descended from simple, diminutive animals that might be thought of as resembling small rodents, though not all possessed front teeth specialized for gnawing. Among the nearly twenty new orders were the one that contains large carnivorous animals, including modern lions, wolves, and bears; the one that comprises horses and rhinos; and the one that includes deer, pigs, antelopes, and sheep. Most of the orders evolved in even less than twelve million years."[151] This kind of morphological change is not only dramatic; it is inconceivable. If we progress from a small rodent to a large rhino in twelve million years, surely there should be some intermediate forms along the way. How is it possible to exchange paws and claws for hoofs in that short time? Stanley foolishly maintains, "The fossil record indicates that whales evolved from small terrestrial animals during, at the most, twelve million years."[152] That is quite a transition to make: small rodent to large ocean-dwelling mammal. This kind of progress is punctuation with a vengeance. Can any rational mind blithely accept this fairy tale as accurate?

There is a Disneyland quality to some evolutionary explanations of change that simply must be noted. Some evolutionists go so far as to attribute sophisticated and careful planning to certain creatures and suggest that their adaptations to the environment were deliberate and not simply chance occurrences. These absurd examples jump out at the careful reader. We will examine some of the more flagrant examples so that we can spot future flights of fantasy of the part of the evolutionists. Robert Bakker presents our first example in his discussion of the ostrich: "Ostriches are long-necked ground feeders, but they have very different problems—they were very long-legged and required their long necks just to reach the ground."[153] But couldn't they just as well have been long-necked and developed their legs to elevate them so they could eat higher vegetation or even so they could see farther to avoid predators?

"Snakes," Bakker tells us, "cannot chew. The evolutionary path they *chose* early in their career required unusual adaptations for swallowing huge hunks of food."[154] (Emphasis added.) How, exactly, did snakes "choose" to swallow instead of chew? Was this a group decision or an individual one? Why make the adaptation if swallowing was so difficult? Bakker has even more thoughtful creatures to describe: "In place of the Late Jurassic's tall browsers, the Cretaceous concentrated on munching close to the ground. Low shrubs and seedlings now faced a threat from herbivores magnified many times compared to the conditions prevalent during the Jurassic. In this ecological context the very first flowering plants appeared on the Earth's surface. How did the Ur-angiosperms react to all this munching close to the ground? One ideal adaptive strategy would be to grow as fast as possible to achieve a height where the lower browsers could not threaten [them]."[155] Lower browsers have been with us since the origin of browsers, are usually miniature versions of the adult browsers, and must necessarily feed on vegetation that is closer to the ground, no doubt thwarting the efforts of the plants. So how or when did the plants decide that enough was enough and begin to frantically grow taller?

We can cite more fantastic scenarios. Bakker tells us that with regard to stomachs, there is a variation in location for the fermenting site. "Ruminants—the deer-cattle-antelope family—*chose* a forward site and remodeled their stomach into a complex multichambered rumen where the bolus is soaked by enzymes."[156] (Emphasis added.) Exactly how did this choice come about? Was there a gathering of deer, cattle, and antelope at which time it was decided to change or remodel stomachs to a new configuration? The fantastical quality of this apparent explanation is precisely what evolutionists themselves complain about regarding the creationists—"Just So Stories" replacing science. There is no question that

Bakker's explanation of the change of stomachs suggests a deliberate act of an intelligent species. Do animals make these kinds of decisions? We can wonder why creationists, who attribute no personality or intelligence to other creatures, have not raised Holy Ned about this use of language.

Steven Stanley gives my favorite example: "In both the bulldog and the panda, the sacrifice of strength in the hindquarters has become tolerable. The bulldog is well cared for in domesticity and is not required to chase down prey."[157] The bulldog must be an amazing creature indeed. Millions of years ago he had a vision of the future, in which he would become the pet of the English middle class, and images of Mary Poppins and the typical English household flashed through his mind. He then decided not to develop strong hind legs because he realized that people would provide for him in a proper way and that he would not have to hunt for food once the English achieved economic and political supremacy. But what about the cocker spaniels and poodles? Well, they didn't trust the English, so they went ahead and developed strong hind legs.

Another important concept in evolutionary theory is that of the "ecological niche," which involves a group of creatures making similar kinds of profound genetic "choices." This idea suggests that there is a variety of body sizes and/or environments that must be "filled" with some kind of fauna during each geologic age. In theory, if a niche is "full," no later organism can grow to that size. The concept of niches seems highly questionable since we are now uncovering a large variety of gigantic dinosaur fossils—pointing to the fact that the large-body niche was filled to overflowing at one time, whereas it is empty now. Adrian Desmond illustrates how evolutionists use the idea of the niche: "Times were indeed dark for mammals. Unable to radiate into large-body niches because these were already monopolized by dinosaurs, mammals were forced to bide their time. It was not spent unproductively, however. During this period they perfected their endothermic physiology by developing an efficient cooling mechanism, involving panting and sweating, and slowly gained an upright stance like the dominant dinosaurs."[158] Is there any proof whatsoever that mammals planned to take advantage of this barrier to growing larger by reorganizing the controls of their body temperatures?

Why couldn't mammals grow larger and compete with the dinosaurs? Little dinosaurs were certainly growing larger to overflow the huge-body niche. How would they know whether a niche was empty or full? Why hasn't any creature attempted to fill the large-body niche abandoned by the dinosaurs after they met their fate in the great Cretaceous-Tertiary (K-T) extinction? Didn't the environment change after the catastrophe that killed off the dinosaurs? If so, wouldn't that change the "niches" that creatures could occupy? Is there no gigantic niche now? Remember that DNA, RNA, and chromosome counts are the primary actors in organic growth. Empty large niches, while tempting, would not be sufficient to allow for the necessary change in DNA so that small rodents could quickly evolve into massive whales. What are these people trying to say?

The most that can be said accurately is that the fossil record, insofar as we have arranged it by first and simplest fossil to largest and most complex, suggests that in Earth history one strange but integrated biotic sphere is succeeded by another equally mysterious. Where organisms came from, whether and how they developed from preceding organisms, and what their fate must have been are all questions that beg new answers. We do not need niches suddenly full or empty to control a species' body growth or cooling mechanisms. Neither evolutionists nor creationists can presently give us adequate answers to these questions.

Evolution, although completely illogical, pervades Western thinking so much that even the best scholars are unable to escape its irresistible allure. Keith Ward, prominent in the movement to reconcile science and religion, explains the conditions that allowed mammals to dominate this part of Earth history. Ward writes that "it is very improbable that self-conscious beings would have ever evolved, and the fact they have evolved on Earth is largely due to the accident that a meteor wiped out the dinosaurs sixty-five million years ago and allowed mammals to develop along a new evolutionary path."[159] Well and good. But two pages later, Ward falls into the seductive trap of orthodox evolutionary doctrines that apparently are not connected to his previous statements in his mind. He says, "The dinosaurs, for example, were selected by the ecosystem as good perceivers and agents, but for various reasons they proved to be an evolutionary dead end."[160] But didn't we just learn from Ward that a massive meteor impact killed the dinosaurs? We have a group of animals well adjusted to their environment and suddenly, out of nowhere, comes a meteor that kills just about everything on the planet. Does this misfortune make the dinosaurs an evolutionary "dead end"? If a piano suddenly fell on Keith Ward, would that indicate that *he* was an evolutionary dead end?

We can see in the confusion in Ward's comment that the task of describing the evolutionary process is at a dead end. At the very least, it is leaving the field of nonfiction. The best course remaining for its proponents is to surrender gracefully and adopt catastrophism in place of uniformitarianism. The political problem we face is simple: (1) catastrophism will aid and comfort the creationists, and (2) catastrophism is the emerging paradigm, and scholars wishing to maintain their status as authorities must now find a way to pretend they have always been catastrophists. That process has already begun. Witness an admission by Stephen Jay Gould in *Science Digest*: "Mass extinctions have been more frequent, more unusual, more intense (in numbers eliminated), and more different (in effect when compared with ordinary extinctions) than we had ever suspected. Any adequate theory of life's history will have to treat them as special controlling events in their own right."[161]

Gould nods politely but doesn't really accept the importance of catastrophes. Others seem more alert. Niles Eldredge presents a much broader perspective: "What we have in the fossil record is a series of more or less devastating disruptions of the world's ecosystems. The less pervasive, those extinctions which affected only some regions, necessarily have left a less dramatic story of annihilation than some of the truly major extinction events, in which scarcely a single species made it across the boundary that is now drawn by the nearly simultaneous disappearances of so many species."[162] Eldredge is no doubt correct in introducing the idea that some catastrophes have been major events affecting the whole planet, while others have been local or regional. Unfortunately, he tries to apply evolutionary whitewash to his insight, contending, "Nearly every burst of evolutionary activity represents a rebound following a devastating episode of extinction. The truly severe extinctions took out up to 90 percent of all species then on the face of the earth."[163] And from this surviving ad hoc group of creatures a wholly new and tightly integrated biosphere with prey-predator and symbiotic relationships arose? Hardly! If 90 percent of the species were extinguished in a major episode, how do we understand Steven Stanley's earlier statement that twenty new orders of mammals evolved in a mere twelve million years from small rodents? That rebound would be truly amazing considering the doctrine of punctuated equilibrium, which asserts that once a species appears in the fossil record, it rarely changes in shape.

We have traveled far enough along the evolutionary road to understand that in its present form, with its illogical dogmas, simplistic reasoning, and fantastic fictional scenarios, evolution has nothing to offer us. Can religion, in any of its denominational forms, offer us anything better? To say that the creatures of a biosphere were "created" forecloses the discussion, so we cannot rely on any Western religious tradition to provide us with a new approach to the interpretation of fossil data. We can find a different framework by which we can view the data, however. Since evolution set out to offer alternative answers to questions posed by Christian religion, it may be that other religions, not dependent on the concept of linear time that is inherent in Christian theology, can shed some light on possible new ways to view the natural world.

CHAPTER FIVE

THE NATURE OF THE PRESENT EARTH HISTORY

EARTH HISTORY IS THE PROVINCE of geology in today's world, often supplemented by paleontology and archaeology. Geological interpretations of physical phenomena on our planet follow a sacred dictum: The processes of physical change we observe today have always been active in the past in the same manner and intensity. Accordingly, we believe we can observe geological events today and project our understanding backward to describe conditions in the planet's life long, long ago. This view achieved dominance in the middle of the nineteenth century as secular science sought to free itself of the restrictive biblical framework. Evolution dictated that life had gradually become more complex over the eons, so the task of the evolutionist and the geologist was to arrange the geological strata according to the complexity of the fossils and search for the "missing links" that would validate the concept of evolution.

This thinking was, of course, a tautology. If strata are arranged in the order of the complexity of their fossils and the subsequent alignment is cited as proof of the validity of evolution, nothing has been gained and no proof has been offered. Most scientists have long since recognized the illogic of this method, but evolutionary biology and geology have maintained their incestuous relationship nevertheless. Derek Ager, until his death the reigning European authority on stratigraphic dating, said it was an impossible circular argument to say that "a particular lithology is synchronous on the evidence of its fossils, and secondly that the fossils are synchronous on the evidence of the lithology."[164]

The rocks are not always found in the proper order; sometimes a formation has simpler fossils on the top and more complex ones on the bottom. To solve this problem, geologists invented the "overthrust." This term is applied to formations in which, in theory, some layers of older strata are pushed up over younger strata, disrupting the sequence of fossils. This concept has been liberally applied when the fossils do not line up in a coherent evolutionary sequence, even when there has been no evidence that strata have been disturbed. Overthrust is thus often used to hide anomalies and explain inconsistencies in the progression of fossils.

We all have seen wall charts in which the various geologic eras, epochs, and periods with their many subdivisions are displayed. We are led to believe that most places in the world provide rocks from each of these respective geological divisions. This lineup is called the "stratigraphic column" and seeks to illustrate the ideal formation in which all geologic times would be represented. But such an ideal arrangement exists only in textbooks. The actual situation is discouraging, to say the least. Alfred de Grazia, historian of science, pointed out, "Rarely does one find even three of the ten geologic periods in their expected consecutive order. Moreover, 42% of Earth's land surface has 3 or less geologic periods present at all; 66% has 5 or less of the 10 present; and only 14% has 8 or more geologic periods represented."[165] Were it not for the tie to evolution, we would have an entirely different Earth history to consider.

The early geologists established the basic length of the respective geologic periods by simple estimates using impossibly naive ideas about sedimentation over time. These periods of time are still sacrosanct for many scientists. Today, with various measuring devices, almost all of which are dependent on the rate of decay of radioactive elements in the strata, we seek to correlate our testing of strata with the geologic ages established arbitrarily in the past. The early assessments of the age of the Earth appear modest now when compared with the current estimates of 4.6 billion years. The problem is that we are simply speculating via computer and are hardly more accurate than people of the early days. In spite of the pious assurances by geologists that the radioactive clocks are accurate, these "clocks" more resemble tarot cards or astrology. There are many problems with current dating procedures. Geologist E. M. Durrance explained, "To obtain the age of formation of a rock or mineral, the material must have remained a closed chemical system since its formation, with neither gain nor loss of radioactive parent or daughter atoms or, in the case of complex decay chains, of intermediate members. Unfortunately, geological material and environments do not often meet this requirement."[166] When, exactly, would the geological material have been a closed chemical system? Of course, the answer is never!

One of the important measuring devices of the early days of geology was the estimation of the rate of sedimentation for certain local areas. These figures were then used to determine the age of a particular stratum by measuring thickness and dividing it by observed annual sedimentation amounts. Thus thick, homogenous strata were credited with immense age, while thinner strata were seen as brief episodes or remnants of a once-thick layer that had eroded away prior to the deposition of newer strata above them. In theory this method sounds reasonable, and since it was obvious that sedimentation occurred horizontally, everyone was pleased with the results. Sedimentation did not last long as a measuring device once the scientists began to measure carefully and do the mathematics, however. Derek Ager demonstrated the absurdities of this method: "A very conservative estimate for the Upper Cretaceous Chalk in northern Europe would give a figure of something like 30,000 feet as an absolute maximum, before consolidation; and about 30 million years for its deposition. That works out as nearly a thousandth of a foot per year, or 200 years to bury a Micraster! And that is for the rapidly accumulating chalk."[167] Obviously, some formations would have to be older than the present estimated age of the Earth if this method were reliable!

The vision of a benign nature carefully laying down its sediments millennium after millennium distracted generations of geologists from observing what was happening in nature. Most did not even take into account the present-day forces of nature that prevented the steady deposition of materials. According to Ager, "The hurricane, the flood or the tsunami may do more in an hour or a day than the ordinary processes of nature have achieved in a thousand years. Given all the millennia we have to play with in the stratigraphic record, we can expect our periodic catastrophes to do all the work we can of them."[168] If sporadic destructive events could wipe away the sediments of millennia in a few hours, how can geologists determine with any accuracy anything about the geologic eras and epochs?

Traditional doctrines suggested that rivers moved debris down to the sea and that as deposits accumulated, the area of deposition obligingly began to sink from the burden in a process called isostacy. The continental shelf thus was capable of receiving more material, and sedimentary deposits of great thickness were made. Simple observation could have told scientists that their scenario was fictional. Most deposits today are riverine delta materials at the edge of the sea that create fan-shaped deposits rather than the extensive and homogenous shales, limestones, sandstones, and the like that we see everywhere. Although some scientists believe that these larger and more homogenous strata occur deeper in the sea, the alleged subduction of plates in modern theory practically precludes these kinds of formations from existing for very long, if indeed they form at all. Ager objected, "Apart from the gradual building out of deltas, with the sediment derived from erosion inland and blown sand moving inland as dunes, nothing seems to be building up along the coastlines or on most of the nearshore shelf."[169] It should have been simple for the geologists to heed the empirical facts and note that everywhere we look, we find none of the processes or sedimentation sites that they declared to exist.

Scientists within the gradualist, uniformitarian school of thought should have asked other simple and obvious questions. Alfred de Grazia queried: "If all this was a very slow process requiring millions upon millions of years, how did it happen that the rivers carried nothing but clay for millions of years and then suddenly changed to sand?"[170] This question is an embarrassment to the purported rationality of science. When I have asked geologists how these changes were made, I have been referred to the "calm, shallow inland sea," of which there are none in existence today. Or I have been told that differential sedimentation is created by wave actions, which would certainly account for some local strata. But these responses do not address the problem of large homogenous strata. Do the calmest, shallowest inland seas sort and separate materials so they can produce sandstones, shales, and limestones? Or do these seas produce whatever geologists need at the time to make their explanation sound scientific?

One geologist made a list for me of the "calm, shallow inland seas": it basically comprised the large lakes of the world and the Red Sea, the Dead Sea, the Great Salt Lake, the Gulf of Mexico, the Amazon delta, the Bering Sea, the Baltic Sea, and the Mediterranean Sea. Sounds impressive—but if one were to examine the possible deposits in the beds of these bodies, one would find that they share hardly anything in common. We would certainly not find uniform beds of limestone, sandstone, mudstone, and clays arranged like many of the formations we see today. As to being calm, the ore ship *Edmund Fitzgerald's* demise should resolve that question; Lake Superior is depositing ships, not limestone.

Even before asking the previous question, we should bear in mind a basic point made by popular science writer Dolph Hooker: "Accumulation, by the agency of running surface water, of any sedimentary deposit, be it rock, gravel, sand, sulphur, gypsum, phosphate or whatnot, necessarily is predicated on the assumption that previous rock masses existed at higher altitudes than those of the present deposits. Definite evidence that such higher rock masses ever existed in many areas on Earth is lacking."[171] In short, orthodox geological doctrines, devised to assist the evolutionists, created a framework that denied the observable facts of the physical Earth.

Alfred Wegener, an early-twentieth-century advocate of the "continental drift" theory, suggested that Earth history must be understood as the breaking up of a supercontinent and the subsequent "floating" of smaller continental landmasses on "plates" over the deeper foundation rocks. After much turmoil, including attacks on Wegener's sanity, geologists have largely accepted this theory today. Although continental drift has solved one problem, it still has not spoken meaningfully to the question of the origins of the vertical columns of sedimentary rock that compose the continents. Presumably the continents have always been above sea level since they "float" on the mantle, while the sea beds are involved in a process of subduction that forces plates under each other in a constant process of renewing the sea floor. Robert S. Boyd, writing in *The Denver Post*, described continental drift this way: "Like a parade of giant tortoises, the plates creep continuously across the Earth's mantle, a 2,000-mile-thick layer of gooey basalt with the consistency of peanut butter."[172] So when do sedimentary rocks form?

Dolph Hooker, in his insightful, long-neglected little book titled *Those Amazing Ice Ages*, posed this important question: "There are thick unadulterated strata of shales, of sandstones, of limestones, of dolomites, etc., covering extensive regional areas. Whether they were separated or indiscriminately mixed in the source rocks, how could such different chemical compounds be separately dissolved, or weathered and eroded, transported, without adulteration, over considerable distances and deposited each by itself in distinct, separate beds and accumulations?"[173] The implications are obvious: We have no good theory to account for any of our sedimentary strata, and that should have been apparent a century ago.

Derek Ager pointed out a difficult technical question concerning sedimentary rocks. As science progressed during the nineteenth and twentieth centuries, scholars from each country tended to name the geological formations they investigated after themselves, their colleagues, or local place names. Few scholars were inclined to investigate the possibility that their strata might extend over extremely large parts of the globe. Nor did they realize that if such extensive strata existed, the complex nomenclature they had adopted would be a major factor in inhibiting the correlation of data. Cutting through the formal nomenclature of national geologies, Ager pointed out that many formations could be found stretched around the globe that were believed to be local, isolated deposits. Among the strata he identified as the same were the following:

1. coal measures from Texas to the Donetz coal basin[174]
2. Devonian red sandstone across northern Europe from Ireland to the Russian Platform, in eastern Canada, and in Kasmir[175]
3. chalk from England, Texas, Arkansas, Mississippi, and western Australia[176]
4. brown building stones in Newark, New Jersey, the Tias of northwest Europe, and the High Atlas of Morocco[177]
5. limestones from Europe, Cantabria, Arizona, the Canadian Rockies, and Alaska[178]
6. banded ironstone in Minnesota, the Transvaal Basin in South Africa, the Hamersly Basin in western Australia, and the Dharwars Series in India[179]

If we group these six major facies that are homogenous and that extend over a substantial portion of the globe, then the theories concerning sedimentation must undergo major renovations. It would be impossible to find a continental source that could deposit these sandstones, shales, ironstones, and chalk according to the traditional doctrines of depositions. These strata compose a significant body of evidence arguing for reconstructing Earth history in an entirely different manner. They may even suggest massive depositions from space that virtually covered the planet in episodes we can scarcely imagine.

The old geology taught that coal and petroleum products were originally composed of organic materials, plants in the case of coal and animals in the case of petroleum. These materials were covered with sediment, subjected to immense pressures, and transformed into the fossil fuels. With coal, an imaginary process of transformation occurred in which vegetative matter became peat, then brown lignite, soft coal, and finally hard coal. But there was no locality in the world where this sequence actually could be seen.[180] The mathematics of coal creation via the orthodox explanation should have alerted people immediately that something was amiss. Immanuel Velikovsky, American psychoanalyst and advocate of catastrophism, pointed out: "It would take a twelve-foot layer of peat deposit to make a layer of coal one foot thick; and twelve feet of peat deposit would require plant remains a hundred and twenty feet high. How tall and thick must a forest be, then, in order to create a seam of coal not one foot thick but fifty?"[181] There was also the problem of explaining the alternating beds of coal and sedimentary rock.

Geologists thought that the ebb and flow of the sea as the area rose and sank allowed sedimentary deposits to be made, followed by the growth and decay of forests to provide the coal strata. But Dolph Hooker pointed out, "In Westphalia, there are 117 beds of coal, one above the other; in South Wales, 100 beds; in Nova Scotia, 76 beds; in Pennsylvania and other coal regions all over the globe there are multiple layers of coal, always with sed-imental strata of clay, shale, limestone, etc., in between. These rock strata vary in thickness from a few inches to many feet."[182] How do we account for this multitude of coal beds? Why, in almost a century and a half, didn't someone ask some incisive questions about the origin of coal? When I asked about this sequence I was simply informed that there was plenty of time for the sea to come and go 117 consecutive times. That would require a stability of climate with extraordinary efficiency to grow 117 forests, interrupted 117 times, anywhere in the world.

It should be apparent that geological doctrines have had a suspect empirical base for a long time. The discovery of unusual formations may lead initially to ad hoc explanations, but after a while, when it becomes obvious that continents are not sinking and rising nor are rivers carrying pure deposits to the sea, some important questions should have been asked and some substantial revisions in theory introduced. It is ironic that the primary effort in interpreting the physical features of the Earth has been simply to claim that present processes, given an infinite amount of time, can produce what we see today. We have nothing except belief to demon-strate the validity of those processes as the solitary factor in shaping the strata of the continents.

Immanuel Velikovsky's revolutionary book *Worlds in Collision*, pub-lished in 1950, suggested the existence of an electric universe and argued that in historical times, human beings had witnessed severe disruptions of the solar system wherein both Mars and Venus had experienced near col-lisions with the Earth. Astronomers, faithfully adhering to the biblical/Newtonian idea of a steady and smoothly functioning cosmos, threw tem-per tantrums of cosmic proportions, going so far as to boycott his pub-lisher, threatening to refuse to publish with its textbook division. It became necessary to transfer the book to a publisher with no textbook line, so intense was the opposition. Although Velikovsky's purpose was to provide a scenario in which Near Eastern chronologies could be reconciled with the Old Testament, the implications of his ideas were staggering. Not only were the heavens again a source of unexpected catastrophe, Western orthodoxy was threatened by the possibility that non-Western oral traditions about a succession of world eras, each ended by major catastrophes, might be reliable. Geology was confronted with the possibility that massive catastrophes had occurred in the past, calling into question the validity of the accepted stratigraphic column.

To bolster his thesis regarding the catastrophic nature of Earth history, Velikovsky, in 1955, published *Earth in Upheaval*, a volume covering a wide variety of data taken solely from geological records and studies. He deliberately excluded the oral traditions and folklore that had characterized his earlier volume in an effort to show that the geologic record alone supported his thesis. His chapter on cataclysmic evolution demonstrated that he was well within the mainstream regarding the origins of life and sought only the recognition that the uniformitarian interpretation of geological data was inadequate to explain the features we find on Earth. Since the astronomers had already blackened his name almost beyond recall, geologists found no need to attack *Earth in Upheaval*, which, based on then-current research, was nearly invulnerable to attack anyway. Some geologists merely echoed the slurs of the astronomers and refused to read the book. *Earth in Upheaval* remains a devastating critique of uniformitarian geology whose questions have never been answered.

On July 16–22, 1994, scientists were treated to a unique event in the heavens. During that period of time, more than twenty fragments of the comet Shoemaker-Levy 9 smashed into Jupiter's southern hemisphere, creating an immense light show as they exploded and flashed when they entered Jupiter's stratosphere and produced great plumes of material in the Jovian atmosphere when they hit its surface. According to predictions, the shock waves produced by the impacts penetrated the interior of Jupiter "at the speed of sound and produced something analogous to earthquakes."[183] Sacrosanct astronomical doctrine of the constant heavens had now been voided by an actual observable catastrophic event. The solar system could no longer be regarded as an infallible machine, and Earth began to look like a small cosmic target rather than the residence of evolution's most cherished product. It is puzzling that the comet Shoemaker-Levy 9 was such a surprise. Derek Allan and J. Bernard Delair, in their book *When the Earth Nearly Died*, point out: "Lexell's comet of 1770 and Brook's comet of 1889 both actually passed through the Jovian satellite system and almost grazed the surface of Jupiter and split in two."[184] Having witnessed a near collision, did astronomers think there would never be any direct hits on a planet?

The adjustment to a solar system vulnerable to massive disruption by outside celestial bodies was not difficult for many astronomers. The idea had been coming anyway. In 1982 Victor Clube and Bill Napier, British scientists, published *The Cosmic Serpent*, suggesting an early cosmic event that was very close to Velikovsky's original solar system scenario. They were well-regarded scholars, so when they announced that a comet might have played a role in Earth and human history, their version of cometary catastrophes was given some consideration. Walter Alvarez, professor at the University of California at Berkeley, had also published several articles on a possible cometary collision in the Yucatan based on the discovery of a thin line of strata that marked the Cretaceous-Tertiary (also called K-T) boundary found in Italian strata, arguing that the dinosaurs might well have been eliminated in the aftermath of such an impact. In 1997 Alvarez published *T. rex and the Crater of Doom* to fully explain his thesis.

Many scientists had accepted the idea that catastrophic events often "closed" geologic periods before Alvarez provided them with reasonable proof. Clube and Napier suggested that the Mesozoic and Cretaceous had been abruptly terminated sixty-five million years ago by a catastrophic event, and various writers began to suggest scenarios that would describe the nature of these catastrophes and how they changed the world around them. David Kring, astronomer at the University of Arizona, provided a modern description of the forces involved in a collision with a sizable celestial body in an article in *GSA Today* describing the impact of the Yucatan comet: "The air blast, for example, flattened any forests within a 1,000–2,000 km diameter region, which would have included the highlands of Chiapas, central Mexico, and the Gulf states of the United States. Tsunamis also radiated across the Gulf of Mexico basin, producing reworked or unusually high energy sediments along the latest Cretaceous coastline. Tsunamis were 100–300 m high as they crashed into the Gulf coast and ripped up sea floor sediments down to depths of 500 m. The backwash of these waves was tremendous, depositing forest debris in 400–500 m of water."[185] We have here an obvious source of power that could create all kinds of geological change, including the deposition or elimination of hundreds of feet of sediment in a very restricted area.

We can hardly imagine the destruction that the Yucatan comet created, and its importance for geology and paleontology can hardly be underestimated. Here we have the explanation of why we find virtual graveyards of fossils in some locations and a glaring absence of fossils in other sites. The back-and-forth motion of tsunami waves must certainly have created strata in which all kinds of materials were buried under alternating deposits of sand and clays. Whether these layers of vegetable and animal matter sandwiched between alternating sedimentary deposits can explain the presence of coal is a topic for future investigation. What should be a certainty, however, is the need to shorten the geological timescale by a substantial percentage. We used to speak of the "rise and fall" of various species, allocating sufficient time for them to evolve and an almost equal time for them to decline. Now we can determine the decline of extinct species in a matter of weeks rather than millions of years if we suspect they were caught in tsunamis caused by a cosmic collision.

The debate now concerns the number of possible catastrophic events that have marked Earth history. Alvarez writes: "The detailed fossil record of the 570 million years since the end of the Precambrian gives evidence of five great mass extinctions and about five smaller ones. The KT boundary is the most recent of the five great extinctions and has yielded much more information than any of the others."[186] Clube and Napier suggest: "With the past 500 million years, then, there have been about fifty collisions of energy more than 7 million megatons, ten of more than 100 million megatons, and one or two of energy in excess of 3 or 4 billion megatons."[187] There are now more than 180 different studies on the subject of comet/meteor/asteroid impacts. Although textbooks continue to reflect a Darwinian perspective on geological processes, the discipline as a whole is rushing toward a belief in catastrophism as the primary agent of geological change.

The support for catastrophism has not been documented clearly since the acceptance of the concept is so recent. Derek Ager, in 1973, wrote: "I maintain that a far more accurate picture of the stratigraphic record is of one long gap with only very occasional sedimentation."[188] Fine, but what does that mean? How do we know there are any gaps at all? Robert H. Dott, in his 1982 presidential address to the Society of Economic Paleontologists and Mineralogists, commented: "I hope I have convinced you that the sedimentary record is largely a record of episodic events rather than being uniformly continuous. My message is that episodicity is the rule, not the exception ... we need to shed those lingering subconscious constraints of old uniformitarian thinking."[189] Yet geological textbooks continue to advocate most of the old ideas. But how can episodicity become a norm until we have some superior timescale to which it can be attached? We can assume some kind of time progression between episodes, but how much? Truly, we have a major task ahead of us to make sense of catastrophic geology.

We can explain some strata as undoubtedly originating in catastrophic events. Thus we can answer Ager's puzzling question regarding deposits in the Jurassic: "One has only to compare the 30 ammonite zones represented in one foot of sediment in Sicily with the 15,000 feet representing a single zone in Oregon to realize how startlingly different rates of deposition must have been in different places."[190] Obviously, Oregon caught hell from a tsunami while Sicily, on the other side of the globe, got off relatively easily. If we accept catastrophic deposition, explaining differences in the thickness of strata is no longer a problem. But what do we do with the horizontal formations and the formations presumably created horizontally that are local but extensive enough to preclude riverine deposits? Or the formations mentioned above that extend over several continents?

In 1958 Dolph Hooker offered a then-novel explanation of the great ice ages, in which he suggested that the water and ice of glaciations had been contained in massive clouds that gave our planet a shroud, similar to those possessed by Neptune, Uranus, Saturn, and Jupiter. But his idea was applied to explain sedimentary rocks as well. "At a recent period in Earth's history," Hooker suggested, "much of the material matter now incorporated in its [the Earth's] stratified crust, as well as a large amount of the water now in its hydrosphere, were still suspended in space above, earlier perhaps in the form of disk-shaped rings and later as overclouding canopies or envelopes extending at times from equator to poles."[191] Strangely, this configuration would conceive of a solar system more uniform than the one we now know in the sense that all planets would initially be endowed with gaseous envelopes containing massive materials that would become the crustal sedimentary rocks.

Hooker developed the idea of the condensation or precipitation of materials onto a planet that was initially incandescent as the solution to the question of the origin of sedimentary rocks. "The atmospheric minerals were more or less separated and segregated into different rings and bands," states Hooker, "which would gravitate and descend to the core at separate times."[192] So we would have a large envelope around the Earth, primarily gaseous, containing the carbonates, silicates, and cosmic dust of various compositions—everything that would come to compose the sedimentary strata we find today, including immense amounts of water.

Then began the process of sedimentation. "As and when the Earth's bands moved poleward and their velocities decreased, eventually the materials in them necessarily fell. Inasmuch as centrifugal force decreased and gravital force increased toward the poles, the materials would tend to fall earlier and more largely toward the poles than toward the equator."[193] Deposits in the northern hemisphere would then be much different from those in the southern hemisphere for each geologic age, depending on the approach of a celestial body and the nature of the material it left in the atmosphere.

We can now answer de Grazia's question about how rivers could carry sand for millions of years and suddenly carry clay or silt. Sand would be deposited before clays and silts that might linger in the atmosphere for a considerable time. Hooker's idea also explains the intense concentration of minerals and ores in some places and the utter lack of minerals in other sites. We simply need sufficient hydrocarbons in the atmosphere falling to Earth to be stratified by the alternating tidal action resulting from the tsunamis to produce the multitude of coal beds found in Westphalia, South Wales, Nova Scotia, and Pennsylvania. Lighter-weight hydrocarbons could fall in a liquid state, percolate through the existing crust, and gather into large subterranean petroleum pools.

The Hooker scenario makes wonderful sense in explaining two bizarre situations cited by Derek Ager. "In the late Carboniferous Coal Measures of Lancashire," he reported, "a fossil tree has been found 38 feet high and still standing in its living position. Sedimentation must therefore have been fast enough to bury the tree and solidify before the tree had time to rot. Similarly, at Gilboa, in New York State, within the deposits of the Devonian Catskill delta, a flash flood (itself an example of a modern catastrophic event) uncovered a whole forest of in situ Devonian trees up to 40 feet high."[194] The Lancashire site is particularly interesting because the sediment to which Ager refers is a coal bed. The sediments can best be understood as a carbon "dump" since the tree did not decompose and become transformed into coal. Forty-foot-high trees buried upright in the ground must have had extraterrestrial assistance since the velocity of water or wind needed to carry forty feet of material would undoubtedly have knocked the tree down.

Hooker's proposal could be adapted to explain the actual formations that we see in the physical world. Initial sedimentary rocks could be laid down in as many layers as we wished and mineral, coal, and oil deposits made at the beginning of a period of planetary existence. Following the evolutionists, we would have a complete biosphere at the beginning of an era. Then an extraterrestrial catastrophe dumps an unimaginable amount of sediment on the planet, virtually extinguishing the existing biosphere by burying it under shales, sandstones, limestones, and other rocks. We might better get these sediments from the action of heat pressure waves, tsunamis, or materials deposited by another celestial body than from the erosion of nonexistent mountain ranges.

Impacts by extraterrestrial bodies can account for many of the features we study, including the greatly distorted mountain ranges that we see today. Continental drift is used to explain these formations, but looking at the twists and turns visible in many locations, it is difficult to conceive of continental pressures causing such destruction. How could formations twist and turn inland, far from the coasts, where the immense pressures caused by drift are allegedly occurring? Mountain chains are the most prominent feature of the continents. Earlier geological textbooks traditionally suggested that there was a time of "mountain-building" at the beginning, during, or at the close of a geologic period, but the actual mechanisms were never clearly explained. A recent textbook, *The Dynamic Earth* by Brian J. Skinner and Stephen C. Porter[195], described the Appalachians, Alps, and Rockies as classic "thrust and fold" mountains although the mechanics, apart from the mysterious continental drift, were never presented clearly. Nothing at all was said about the origin of the Cascades, the Andes, or the Himalayas. Presumably these mountain chains were formed by thrust and fold as well; if not, one wonders how they came to be.

Catastrophism involving extraterrestrial forces offers a much better scenario to explain the presence of mountains on an otherwise relatively calm globe. Donald Patten, geographer and advocate of catastrophism, suggested that mountain building was the product of a close fly-by of a small celestial body that had sufficient speed and gravity to raise mountains like welts on the surface of the Earth. Looking at the surface of the whole planet, he noted: "The mountain systems of the Earth are found in great scallop-like arcs, which in turn merge into greater arcs, which in turn merge into sweeping, planet-traversing circles. Their pattern seemingly is indifferent to either continental massifs, or to oceanic basins; they traverse either with equal ease."[196] These arcs are what constitute the chains of mountains that we identify as distinctive. To credit continental drift with their creation would mean that its force could create scalloplike configurations, a concept that is doubtful and that is ignored in textbooks.

In Patten's theory a smaller body—for example, the planet Mercury—could, as it joined the solar system, substantially disfigure a planet such as ours by two or three close passages that would raise mountainous arcs in a matter of weeks. In fact, Patten argues, "In a few hours of fly-by time, as much 'work' was accomplished deforming and reforming the Earth's crust as evolutionists allow for 200,000,000 years."[197] This mechanism would easily create severely twisted strata in various parts of the world, which would represent large amounts of currently accepted geologic time. Since we have seen that tsunami waves can reconstruct a landscape and severely disrupt ocean-bottom sediments, eliminating deposits that were believed to have taken millions of years to accumulate, a fly-by could easily reduce the geological timescale by a startling amount.

We now have to examine the problems existing within our current orthodox view of Earth history. Many of the old geological concepts must be severely limited in their use as explanations of geological features and formations. We need several theories of sedimentation in view of Ager's list of massive strata that encompass several continents. We need a realistic explanation of the origin of mineral deposits and the creation of the "fossil fuels." Considering the large amount of coal and petroleum we have used in the past several centuries, can we seriously continue to pretend that vegetation and animal remains alone were the source of these products? Were Saudi Arabia, Kuwait, and the Arctic Slope stacked high with animal carcasses in some remote geological past? If so, how did that happen?

Catastrophism enables us to reduce the geological timescale by a substantial margin, bringing formerly remote periods of geologic time much closer to the present. But by how much does the timescale shrink? Here most geologists will reach a state of angry incandescence. Many scholars have spent a good deal of time devising ways of measuring the age of rocks without recourse to the simplistic measuring of the thickness of sediments. They have in fact extended the probable life of our planet into the billions of years with new tests and procedures for measuring geologic time. But there are many real problems with the various measuring devices used today.

Almost all contemporary measuring devices—the potassium-argon and other radioactive trace element tests as well as the carbon 14—have some fatal logical errors that render their claims useless. Measuring a rock involves the assumption that we know the initial conditions under which it was formed and the further assumption that no intervening events occurred between its origin and the time we measure it. No strata on Earth can meet these conditions because we simply do not know the beginning, nor do we know the developmental history of the rock. We know only the relative percentages of minerals or elements contained within the rock today. We gloat over the potassium-argon dating method, but it has severe problems. De Grazia states, "Argon, like uranium and radioactive trace elements generally, tends to rise to the surface of the Earth. Hence surface rocks (and these include all that have been measured) will be high in argon content. Argon also can be infused into hot rocks from the air and kept there as the rocks cool."[198] Given that the Earth has a history of collisions with extraterrestrial objects, each of which would radically change the ratio of radioactive elements in the rocks, measuring surface rocks to determine age is hazardous at best and probably a futile enterprise.

If we are reducing the geological timescale significantly, bringing the remote periods much closer to the present, can we now consider matching the stories, legends, and beliefs of non-Western peoples with either the geologic periods or the catastrophes that ended these periods? If many of these formerly long periods of uniformitarian time can be accounted for in a matter of weeks by a close fly-by, then we must look carefully at the descriptions by ancient peoples of the manner in which geologic periods closed. Ancient peoples described certain periods of Earth history as "worlds" or "suns" or "ages" and remembered that they ended with great catastrophes, such as a rain of fire, a horrendous flood, a tremendous wind, or brimstone and hailstones. Can these descriptions be linked to possible extraterrestrial impact events?

Derek Ager suggested several times that ancient peoples must have experienced some spectacular geological events: "Many early humans must have seen geological phenomena far more violent and spectacular than any we know in historic times, including the last great volcanicity across northern Europe from the Auvergne to Romania and the explosion of Santorini which may have given rise to the Atlantis legend. In New Zealand the first Polynesian immigrants may have seen and suffered some of the last huge volcanic explosions in North Island."[199] He also suggests that "one of the most dramatic sights ever seen by humans must have been the lava flow which cascaded more than 900 m over the edge into the Grand Canyon less than 10,000 years ago after the arrival of the Native Americans."[200] Surely these suggestions invite us to consider how many catastrophes may have been seen and experienced by our ancestors.

Clube and Napier project that "within the last 5,000 years, that is, within a timescale of interest to the archaeologist, the historian, and the mythologist, there must have been about fifty impacts in the energy range of 1–100 megatons, about five in the range of 100–1,000 megatons, with an even chance that there has been an impact in the range of 1,000–10,000 megatons."[201] But suppose that a major fly-by had occurred within the time period of interest to these scholars, raising the Cascades and Andes, shifting the Alps eastward and tilting them, or creating the Himalayas? Much of what we regard as geological knowledge would have to be radically revised in favor of a drastically shortened chronology of Earth history.

Deriving knowledge of Earth history from oral traditions is exceedingly difficult because of the prevalence of the doctrine of uniformity in scientific circles. Traditional stories about the world ending with massive floods and fires are not taken seriously because we have been taught to believe that no forces more powerful than those we can presently observe have ever been active on the planet. And we have been taught that throughout Earth history things have been calm and uniform. Most of our textbooks continue to reflect this calm, orderly progression of life that we know is not accurate. Many scholars continue to interpret data in the old context, incapable of conceiving the revised history of our planet that is unfolding every day.

Now we have the pictures of the comet Shoemaker-Levy 9 hitting Jupiter and wreaking untold damage and change. We understand that we are vulnerable to an unexpected visitor from the sky that could eliminate us in the blink of an eye. In March 1998 a group of astronomers announced that an asteroid would pass within 30,000 miles of the Earth with the possibility of a collision. NASA officials at the Jet Propulsion Laboratory promptly recalculated and said it would miss us by 600,000 miles.[202] Again, on June 15, 1999, the Millennium Group of scientists announced that Comet Lee might collide with us.[203]

Scholars often comment on the frenzy with which ancient peoples observed the sky. Perhaps they had experienced a minor catastrophe and fully realized what might be visited upon them from the skies. We are just as nervous as they once were. A NASA press release of July 22, 1999, announced: "A risk-assessment scale, similar to the Richter scale used for earthquakes, will assign values to the celestial objects moving near the Earth. The scale will run from zero to 10. An object with a value of zero to one will have virtually no chance of causing damage on Earth; a 10 means certain global climatic catastrophe."[204] In the practical, if not the academic, world catastrophism is now the dominant paradigm.

If scientists are willing to accept a period of 5,000 to 10,000 years prior to the present as a time when the accounts of ancient peoples might be true, why is 10,000 years the limit beyond which memories are not valid? Why can't we go far back and link the old stories with geological findings and create a truly secular and neutral history of the Earth? Searching folklore of the remote past to learn of events that might have radically affected the surface of the Earth may appear foolhardy, but it may also provide us with data we had not previously considered. Graham Hancock, a popular British science writer, has explored many sites of probable underseas cities all over the world. This information has not previously been available to us because no one thought to look. His data begs the question of whether seacoasts suddenly dropped or the planet received so much water from outer space that the continental shelves were flooded.

I have not recently met a geologist who does not embrace some form of catastrophism. But the adoption of the idea is akin to having a firearm. If needed to explain something it may be used, but generally the idea is that uniformitarian principles must hold as long as possible. Thus, strata that can be easily explained as the result of extraterrestrial impact will remain within the uniformitarian framework, and the scope and nature of the catastrophe will have to be resolved by the astronomers.

CHAPTER SIX

THE NATURE OF "RELIGION"

AS WE HAVE NOTED, evolutionists and creationists disagree over the mechanism by which organic life originated on this planet. Although we have argued that evolutionary explanations are inadequate (primarily the ex post facto imposition of evolutionary dogma on the fossil record), creationism's answer to the question of origins terminates the inquiry. Once you have said that god made everything, nothing more can be said. Like "evolved," the label of "created" will adorn every experiment's results and serve as the obligatory preamble to scholarly reports. There will be no need for additional discussion. That situation is hardly comforting to the millions, perhaps billions, of people who want some kind of answer to the question of origins. What can the religious traditions provide that will help resolve the present impasse? What is religion that we look so expectantly toward it for answers?

Religion has a status like evolution in our minds and emotions. As with evolution, the content of religion is defined by one's personal preferences, so that almost any belief or activity can be described as religious. There are two approaches in the modern world to conceiving and understanding religion: reductionism and relativity. In reductionism, as Ian Barbour explains, "Religion is just psychology, psychology is basically biology, biology is the chemistry of large molecules, whose atoms obey the laws of physics, which will ultimately account for everything."[205] Here religion is a minor phenomenon to be studied by psychologists, and not mainstream psychologists at that. Religion has come to be seen by many as a mild form of mental illness or immaturity to be transcended by science in its ability to give us answers.

Some scientists sincerely believe that science must eventually triumph. Frank Tipler, a professor of mathematical physics at Tulane University, in *The Physics of Immortality,* argues, "The universe is defined to be the totality of all that exists, the totality or reality. Thus, by definition, if God exists, He/She is either the universe or part of it. The goal of physics is understanding the ultimate nature of reality. If God is real, physicists will eventually find Him/Her."[206] One of the hot topics today in physics is the search for the Higgs boson, a particle nicknamed the "god particle" because it would help explain some complex equations.

Of course, the answer is not as simple as finding a particle. Tipler's comments do, however, illustrate the propensity of scientists to propose a concept that unites the various theories about the world in some easily expressed principle and call it "god," contending thereby that science can provide all the answers. But the personal relationship to god that many people feel they have will not be explained by this conclusion. The devotional life of millions and the unexplained revelatory experiences of thousands will continue as before. Nothing will have been explained or gained.

The contemporary view of the relativity of religion is a product of nine-teenth-century scholarship that began to subject sacred texts to the same kind of scrutiny that had been applied to other ancient literature such as *The Iliad* and *The Odyssey*. Various Eastern religions were encountered and studied, and their sacred books were translated into European languages. Archaeologists began making available religious texts of extinct societies while anthropologists frantically recorded the religious traditions of tribal peoples before those people became extinct and were forgotten. Many of these traditions expressed sentiments similar to those of Christianity, and a sense of commonality among various religious traditions emerged.

In Germany in the nineteenth century there was a great effort to estab-lish in a scholarly fashion the probable life of the historical Jesus. This movement crested abruptly when Albert Schweitzer published *The Quest for the Historical Jesus,* which sketched a realistic portrait of the man as a deluded revolutionary hero. He followed this book with a psychoanalytic study of some depth that proved too realistic for most Christian theolo-gians. Recent studies of the Dead Sea Scrolls have revealed evidence com-patible with Schweitzer's scenario, further undercutting the sense of the supernatural that Christianity once possessed. The Christian religion, forced to deal with historical realities, was seen by many thinkers as sim-ply another belief system from the past, one that transcended competing religions at the time of its origin and that also emphasized dying saviors.

Virtually banished from the evolutionists' discussion about the many discoveries of the latter half of the nineteenth century, mainstream American Protestantism withdrew from active resistance to Darwinism. The social gospel movement in America, best personified by Baptist min-ister Walter Rauschenbusch and later by Reinhold Niebuhr, a systematic theology professor at the prestigious Union Theological Seminary in New York, emphasized the responsibility of Christians to improve the social and economic conditions of the world. This movement uncritically accepted the accomplishments of science and ultimately reduced religion to a command to do good works. Some people even regarded the religious task as one of building heaven on Earth using scientific discoveries and technology. The sense of awe that had always been an essential part of religion was lost, and reasonable propositions about the world replaced it.

Fundamentalist Christians soundly rejected evolution and took refuge in the comfortable belief that the Bible contained the literal truth because it was the word of god. They simply did not care to debate the question of origins since the Bible was quite clear on what had happened and how it was done. This attitude led to the Scopes trial in Tennessee in the 1920s, which addressed whether evolution could be taught in public schools, as it then was prohibited by state law. Creation thus occupied the role and status now accorded evolution. The trial was probably the high-water mark of Christian influence on public education, and the ridicule heaped upon creationists afterward elevated evolution to superior status in the classrooms. The Roman Catholics, of course, had to wait for a papal pronouncement that was a long time coming. American Christendom consequently is badly fractured between evolutionists and creationists, a division that certainly suggests that evolution has strong religious overtones.

Langdon Gilkey, a liberal Protestant theologian, described how the concept of religion was transformed by the critical studies of secular scholars: "Religious truths were considered to be relative to their culture, and therefore subject to change, development, and, in principle, at least, irrelevance and abandonment."[207] All the religious traditions were now understood as ancient fiction created to educate children in the ways of their societies. When addressing the rest of the world, liberal Western thinkers granted some measure of equality to other religious traditions. In a further surrender to science, the practice of Western thinkers was to regard non-Western religious traditions as examples of cultural evolutionary processes, each religion being a stepping-stone marking man's way. Following the scientific example, the best religion, then, was one that formulated the most sublime concepts to describe the deity. Some early-twentieth-century writing in fact suggested that the purity of the formulation of the concept of deity distinguished Christianity from lesser religions.

This practice of placing Christianity at the apex of religious development, with other religions not quite at its level of understanding, continues today. Ian Barbour, in suggesting a new way of solving the present science/religion conflict, wrote: "We can therefore tell an overarching story that includes within it the evolution of life and human beings, continuing in the stories of covenant and Christ—*with a place in it for the stories of other religious traditions.*"[208] (Emphasis added.) Exactly where would other religious stories be placed? When they happened to agree with Christianity, they would be found in the footnotes and anecdotal references. When they did not agree, they would be cited in a pejorative context to emphasize the superiority of Christianity.

In order to tell a story that would make sense to scientific minds, we would have to formulate a new concept of religion. We have been raised within the confines of the nineteenth-century view of religion, in which our species was believed to have evolved from a simple belief in animism to the lofty reaches of Protestant theology characterized by American theologian Paul Tillich's "god above god." This outmoded format must be surrendered and a new perspective advanced that will give religion its due. Arnold Toynbee, noted British historian, sought to articulate a different perspective that would give the other major world religions some respectable status. "There is," he wrote, "a possible psychological explanation of the survival of eight faiths on one planet. It is possible that each of the surviving faiths may prove to have an affinity with one of the diverse alternative possible organizations and orientations of the Human Psyche."[209] Here history and psychology are combined to validate and rescue religion.

This solution would be satisfactory to many scientists, but it overlooks one basic fact. A religious tradition becomes a "world" religion primarily by the size of the population of its adherents; it does not necessarily represent a different psychological approach to understanding the world. Judaism, Islam, and Christianity actually represent but one psychological tradition separated by their arguments over historical facts. Hinduism, Buddhism, and other Eastern religions represent another psychological tradition separated by many devotional methodologies. Then we have the thousands of smaller shamanistic religions that are so attractive to people today, with a psychology that emphasizes experience over doctrine. We do not really have eight distinctive psychological paths at all but several general approaches.

People grow up and live in different social contexts where particular beliefs are assumed to be valid and so are rarely questioned. They do not choose from among the many different possible answers to ultimate questions but simply assume that the worldview in which they have been raised is the proper one. Once a particular religious worldview is accepted by a society, even missionary efforts to change that context have little effect— except when conversion is accomplished by force. Europeans, it must be noted, were often converted by force and thus became monotheists worshipping abstract conceptions of the deity in order to survive.

The biggest problem we face is our failure to understand that science and religion, as distinct concepts, are terms used primarily in the Western intellectual tradition. They came into use when the West accepted the division of knowledge into the realms of reason and revelation in the late Middle Ages. Western thinkers since then have assumed that all societies they encountered had also split their knowledge between sacred and secular, rational and mystical. In the centuries of rapid colonization, whenever one of the European countries encountered other peoples, they were surprised at their lack of technical scientific knowledge. Consequently, they decided that everything other cultures knew or believed that did not explain or promote the scientific viewpoint had religious connotations. Further, because of the Europeans' firm belief in cultural evolution, non-Western beliefs were regarded as superstitions.

Western science, when it encountered information from other cultural traditions that was arranged in a different format, rejected any knowledge that did not fit into its cause-and-effect analysis of the world. As Western science became more successful and adopted the Newtonian mechanical universe, much knowledge from other cultures was lost because it could not be explained in Western terms. The same reasoning applied to religion. In Western history the original motivation to conquer the world came from missionaries who, following the commands of the Acts of the Apostles, desired to bring the gospel to all peoples. The religious questions and answers posed by Christianity therefore were seen as the only way to describe religion. Since most other religions did not have such concepts as grace, sin, redemption, or salvation, they were understood as incomplete and, unfortunately, in need of Christian reformation.

Beliefs and practices of other peoples had to fit into the familiar Christian format of a single creation, the Fall, sin, repentance, redemption, salvation, and judgment. Students of the history of religion, in addition to locating other religions further down the evolutionary scale than the Western religions, tended to reshape those religious traditions and interpret them in Western terms. Often, to gain an advantage in proselytizing, efforts were made to identify certain aspects of Christianity within the beliefs and practices of other peoples. When one set of identities was made, the other beliefs and practices not consonant with Christian teachings were believed to be shortcomings, thus elevating Christianity above these traditions and reinforcing the misunderstanding.

Most tribal people, including American Indians, talked about a great mysterious energy that they believed undergirded all life activities and was to be found everywhere. They used various terms to indicate how this energy was experienced and perceived on different occasions, but missionaries and scholars forced them to choose a single term that could be translated and stand for "God the Father." For example, of the many words in the Sioux language that expressed a wide variety of experiences and apprehensions of a universal, superior personal intelligence, the Sioux were forced to choose "Wakan Tanka." Their original conception of this energy was transformed into a belief in a "god" that, in practice, resembled the god of the Old Testament. Thus Father Schmidt, noted German scholar of the history of religions, spent his life searching for the concept of a "high god" among the North American Indians, finally weakly arguing that this concept was found primarily in northern California.

Once we step outside the Western framework, we realize that religion is not a universal concept and is probably not even a proper category to use when examining human cultures and their perspectives on the world. Murray Wax, sociologist of religion, among other scholars, asserts that "only the languages of the modern West contain a term corresponding to religion."[210] Tracing the concept backward in time, Wax found that "even the ancient Hebrew Scriptures contain no word corresponding to religion."[211] Except in the modern West, then, there are virtually no "religious" traditions that separate religion from the rest of human knowledge and experience.

Some traditions are immensely practical and pragmatic. Instead of a fragmented worldview in which science and religion are opposed to each other, as we find in the West, these traditions speak of ways of living and how our species can best deal with the life experience. Their teachings and practices are a mixture of insights derived over long periods of time. Their primary concern is whether or not something works. A judicial scenario of sin, judgment, redemption, and so forth does not burden them. They do not even use the same words to describe their activities. Keith Ward comments, "In the teaching of the Buddha there is outlined a path to that wisdom, compassion and bliss that delivers from evil and suffering. In

Confucian teaching, the 'way of heaven' is a way of justice, order and the harmony of all things."[212] Are these philosophies, religions, or even psychologies?

Western thinkers know full well that none of these other traditions represents anything remotely approaching the idea of "religion" as we conceive it in the West. Ian Barbour admits that "some scholars have studied diverse cultures and concluded that religious traditions are *ways of life* that are primarily practical and normative."[213] If a tradition is seen as a way of life that includes and explains all phenomena, not a separate and distinguishable activity of a society, how can we have a universal idea of "religion"? In other words, what we generally refer to as "religion" is, in other cultures, simply a comprehensive worldview that contains ideas that in the West would constitute separate academic disciplines. Instead of scientific "facts" versus religious sentiments, these people have an ensemble vision of the world around them and seek to find our species a harmonious place in it.

This attitude is wholly empirical, as opposed to the dogmas and abstractions of the West. There are many examples of this attitude toward the world, and one taken from the American Indians may illustrate the problem of separating beliefs into subcategories. In 1918 A. McG. Beede, a Christian missionary on the Standing Rock Sioux reservation in North Dakota, published a manuscript on western Sioux cosmology in which he recounted conversations with the elders of that tribe. "There is no difficulty in leading an old Teton Sioux Indian to understand the 'scientific attitude,'" he wrote, "that the processes that give rise to phenomena may be more and more known by man and may be, to some extent, controlled by man, and that in this way the forces of nature may become a mainspring of progress in the individual and in the human race. The idea of atoms and electrons is easy and pleasing to an old Indian, and he grasps the idea of chemistry."[214]

Beede then took Harry Boise, a Yale graduate, with him to Turtle Mountain to instruct the Chippewa and Cree Indians there about science, as he had done with the Sioux at Standing Rock. Boise made his presentation and got the same reception as Beede had at Standing Rock; the American Indians had readily absorbed the scientific ideas he presented. Boise reported that after his presentation, however, a Chippewa leader, Rising Sun, rose and gave the response of the group: "Rising Sun, speaking the conclusion of all, pronounced 'the scientific view' inadequate. Not bad, or untrue, but inadequate to explain, among many other things, how man is to find and know a road along which he wishes and choose to make this said progress unless the great Manitoo by his spirit guides the mind of man, keeping human beings just and generous and hospitable."[215] The Indians clearly placed moral and ethical knowledge above any knowledge of the physical world in their more comprehensive framework of understanding.

Beede then remarked that he had been told by Red Thomahawk, a Sioux elder from the Standing Rock Reservation in North Dakota, that "the knowledge and use of any or all the powers of the objects on Earth around us is as liable to lead a man wrong as to lead him right. It is merely power, with no way of knowing how to use it correctly ... unless Woniya [Spirit] is with a man's spirit for the light. This is the old teaching. It is true."[216] Perhaps the other traditions that we have described as religious have also been able to understand the factual physical things of the world but have been more interested in learning how best to live their lives. Maybe they were not interested in achieving power over nature, a power that they considered to be illusory.

Was Toynbee completely wrong in suggesting that he could find "religion" present in eight traditions that were in fact important psychological insights? Not exactly. Preliminary examination of these eight traditions would reveal differences in belief and practice. More extended contact between traditions would perhaps encourage competition, and the number of converts would be cited to prove the superiority of some of the traditions to the others. But prolonged contact, such as we have experienced in the past century, would involve a mixing of worldviews and psychologies and would provide the groundwork for examining the "religious" question in a more profound manner. Thus we have books arguing that the Tao can be seen in quantum physics or, as in Gary Zukav's *The Dancing Wu Li Masters*, that the Asian traditions explain the physical process of the universe in spiritual language.

Hans Küng, noted German Catholic theologian, has suggested that indeed we face a unique situation in religion today: "After the discovery of the giant continents outside of Europe the world religions were first and foremost an external *quantitative challenge* for Christendom," he says. "But they have now become an internal, qualitative challenge not just for some enlightened spirits but for the Christian churches themselves."[217] The original challenge faced by Christianity was the discovery of tremendous numbers of non-Christian peoples on other parts of the planet whose existence required an explanation. Now the challenge goes to the substance of religion. Robert Bellah, sociologist of religion at the University of California at Berkeley, points out that "the symbolization of man's relation to the ultimate conditions of his existence is no longer the monopoly of any groups explicitly labeled religious."[218] If other insights are now competing directly with the Western religions, and they certainly are, why aren't these beliefs being integrated into the science/religion debates? More important, how can the controversy in Middle America's schools and courtrooms be seen as a science/religion conflict when it is only a quarrel between science and fundamentalist Christianity?

One of the major problems in integrating other traditions into the discussion is that many of them are based on a simple recognition of the intelligibility of the natural world. This includes both small, isolated tribes of people in relationships with spirits and animals in their own environment and Hinduism and Buddhism, which confront the larger cosmic rhythms. No one can deny that on the practical, experiential front, this tradition of naturalism has made great inroads into the institutional religions of the West. Although couched in a New Age format and sometimes presented almost cafeteria-style, these beliefs and practices are quickly replacing traditional Western interpretations of the world.

G. Van Der Leeuw, the great German professor of religion at the University of Groningen in the 1930s, suggested that the same misuse of concepts has occurred with respect to nature. He observes, "What we moderns call 'nature,' in fact has a prominent role in all religions without exception. Yet it is neither Nature nor natural phenomena as such, that is ever worshipped, but always the Power within or behind."[219] But when other traditions were placed within the evolutionary framework, they were characterized as animistic, as the superstitious worship of nature—a stage that Western peoples believed they had transcended. In fact, the other traditions were simply reporting their experiences in the world and giving interpretations of what they had experienced. For them, there was no additional incline to ascend. The "purity" of the concepts used to describe religious experience was irrelevant to them.

Strangely, Christian theologians continue to insist that evolutionary interpretations of religion are the only way to understand the problem of religion. Thus Ian Barbour pleads for the inclusion of nature in theology but connects this inclusion to the observations of science—which are changing daily. Barbour writes, "We seek a theology of nature. Such a theology must take the findings of science into account when it considers the relation of God and man to nature, even though it derives its fundamental ideas elsewhere."[220] Would not a "theology of nature" take nature as it is without importing analytical concepts from outside, however? Since scientific findings are always changing, how would a theologian determine the proper scientific expression to which he could respond?

Barbour himself illustrates some of the difficulties in connecting theology and scientific discoveries. Noting that contemporary theologians do not adequately deal with nature, Barbour admits: "Discussions of providence, for example, refer extensively to God's activity in nature. What, then, was God's role in the long stretches of cosmic time before man's appearance? Is a sharp distinction between history and nature tenable, if nature itself has a history and if man is rooted in nature?"[221] What happens to theology if god waits for eons for an evolutionary process to produce an intelligent ape? Wouldn't that condition again return us to a wholly naturalistic theology absent the major doctrines of Christianity? Could we rely on a religion that appeared only in the recent blink of a cosmic eye?

While many of the evolutionary explanations are untenable, we should have no quarrel with the proposition that our planet is very old and that there has been a considerable natural history prior to the appearance of modern human beings. If, however, we are the result of a random meteor hit that eliminated the dinosaurs and allowed mammals to become the dominant form of life on the planet, was this scenario a part of "god's plan"? If so, we can only conclude that the meteor hit was a deliberate act of god and not an accident. And we have made no progress whatsoever, attributing divine intervention to natural processes, and raising serious questions about "god's will." Merging science and Western religion, even on the most flexible terms suggested by liberal Christian thinkers, simply does not work. Theology cannot simply insert itself at convenient spots in the scientific story.

We like to believe that the histories created by Western science and religions are the proper way to understand the world. They do not, however, produce the results claimed by them. Once comfortably placed within a historical context, be it evolutionary dogma or a salvation history, people tend to believe they are in control of things. They are no longer in awe of anything except themselves. A direct and continuing relationship with nature has far different results. Ian Barbour admits: "Personal responses to the sacred in nature are indeed *more universal and less divisive* than particular theological doctrines in an age of religious pluralism. They encourage humility, and openness, avoiding the dogmatism that has often been present in historical religious traditions."[222] So the focus on history must create dogmatism and divisiveness. Particular histories often highlight such ideas as the "chosen people," "manifest destiny," and "1,000-year Reichs," which are not beneficial to anyone.

If we have defined the non-Western religions as cultural worldviews that are based on an empirical knowledge of nature and are lacking the division between science and religion that we find in the West, are there other problems in defining religion? Western thinkers spend an inordinate amount of time locating "god" within their schemes. But this approach is generally a "stick-'em notes" activity: attaching "god" to whatever twists and turns in the cosmic process are useful. Some of these efforts are indeed clumsy. Keith Ward suggests: "So far as science goes, God could suspend natural laws occasionally, for a good enough reason."[223] What? Or we can look at the comment of W. Mark Richardson of the Center for Theology and the Natural Sciences, as reported in *Newsweek*: "Science may not serve as an eyewitness of God the Creator, but it can serve as a character witness."[224] Neither scientists nor theologians should take these comments seriously. We need simply ask exactly what kind of "god" is being inserted into otherwise neutral scientific accounts—and whether he/she *needs* a character witness.

We are accustomed to jumping from presuppositions to unwarranted conclusions, and nowhere is this tendency more pronounced than when we speak of "god." Van Der Leeuw contends that "it is a failing of modern thought that, in connection with the term religion, it must immediately think of 'gods.'"[225] And, he tells us, "'god' can be applied only to what a modern Western European, descended from the Christianity of the age of 'Enlightenment,' is accustomed to designate by this name without further philosophical or phenomenological reflection."²²⁶ So even the concept of god is a wholly modern European concept that has come to dominate discussions about religious phenomena even though it derives from but one tradition on the planet and is not shared by other traditions.

Eric Sharpe, biographer of Nathan Söderblom, the great Swedish theologian, commenting on Söderblom's work in the history of religion, explains our problem well: "The attaching of undue importance to the conception of divinity has often led to the exclusion from the realm of religion of (1) phenomena at the primitive stage, as being magic, though they are characteristically religious; and of (2) Buddhism and other higher forms of salvation and piety which do not involve a belief in God."²²⁷ If the study of religion involves analysis of behavior and not merely the concept of "god"— involves what people *do* and not just what they *say*—then the arguments of American creationists fall completely apart. The perspective of the so-called primitive peoples becomes considerably more important for our study, and our thinking moves to a more practical/emotional foundation.

Keith Ward describes an important characteristic of Eastern traditions that lack the concept of deity in the Western sense: "The Eastern stream in which Buddhism, Taoism, and Confucianism interact develops from forms of animism to the idea of a cosmic order, a way of balance and harmony following which brings stability and calm of mind, and peace and right order in society. In this stream, there is little stress on one Absolute Being or God."²²⁸ But isn't this state of calm of mind and peace and right order in society exactly what Christians claim for the individuals who have accepted their religion as a gift from god? Where do we get the idea that there must be a god who has a "personal" relationship with each individual human? It does not come from the observation of the natural world or experiences in it. The religions with a god or gods must be relying on their own historical experiences. The task is to determine what those experiences might have been and when they might have occurred.

The corollary inquiry about the necessity for a god is whether or not the non-Christian religions have a "creator." In the other traditions the concept of the creator is useful for answering the question "Where did it all come from?"—but for little else. The creator in many religious traditions is separate from creation, and in many instances, once having completed his task, wants nothing more to do with his creation. In these traditions, at least in everyday life, there is not much speculation about the creator, his attributes, or his role in planetary events—though there is a sense of basic appreciation. Great catastrophes, which shake our faith in the regularity of nature, do seem to invoke a momentary appeal to a deity or creator. But the argument citing this momentary response to a crisis as evidence of divine activity, or even divine existence, falls apart in the normal course of events. Faced with insuperable odds, we naturally look for powerful assistance to make things right. Having believed in the regularity of nature, we find it unfair that catastrophic events might injure us, refusing to recognize the larger cosmic processes to which our planet is subject. But in our daily lives, for the most part, the creator plays no significant role in our decision-making, behavior, or attitude.

Hans Küng supports this interpretation: "It seems indeed possible to prove the existence of primitive tribes who believed not in spirits but in a 'high god' (primordial or universal Father as father of the tribe or of heaven), although the latter—oddly enough—[has] little or no place in worship and apparently functioned merely as 'originator' to provide answers to the questions about the source of things."[229] Important to note here is that while the creator is given credit for originating the phenomenal world, the focus of the tribal peoples is on those relationships with spirits that can be beneficial to them. The spirits may themselves originate from the creator, but they are always the active agents in people's lives. The creator, in fact, appears to be something akin to Aristotle's "prime mover"—he begins the universal process but is not seen afterward except in logical syllogisms intended to prove that the creation was designed and can be rationally understood.

When we look at the major Eastern religions, we find that they also do not link the creator, if indeed any is articulated, with the activities that Western thinkers call "religious." Ernest Benz, a history of religions scholar, says that "Hinduism, like Buddhism and Shintoism, lack one other distinction so fundamental for our Christian thinking: the belief in the basic essential difference between creation and Creator."[230] He suggests that "the same central importance that the idea of the absolute otherness of Creator and creation has for us, the idea of the unity of being has within Buddhist and Shinto thought."[231] The West, in adopting the posture that the creator is continuously linked with our daily lives, must necessarily feel threatened when science offers a different interpretation of things. But it would be folly for us to link our conception of the creator with any expression of science. So while American school boards struggle with evolutionists over the existence of a creator, people in many other cultures have no difficulty discerning cosmic rhythms and adjusting their lives to be in harmony with them.

According to Langdon Gilkey, even Christians cannot talk about the creator *except* as they discuss the natural world: "In the Jewish and Christian traditions it is generally accepted that God, whatever the divine nature may be in itself, is known and spoken about by us as God reveals his nature and purposes through creation, and especially through history and persons. *We have difficulty*, in these traditions, experiencing and speaking of God in himself. Thus we speak of God *in conjunction with* that aspect of the creaturely world—be it nature, history as a whole, or special events and persons or communities—in which God is present and through which God is manifesting himself."[232] (Emphasis added.) Basically Gilkey is admitting that *no* religious tradition can make significant statements except as they relate to the empirical world, and this limitation is precisely the approach of the tribal peoples. The argument can be made that this approach is merely the projection of personal psychological hopes and beliefs. The fact remains, however, that people in other traditions can often perform feats that cannot be duplicated by either theologians or scientists in our society.

Can we live in and observe the natural world without positing the existence of a creator? The traditional logic states that from observations of the world, there seems to be an orderly progression of events, and everything seems to be coordinated with everything else. Thus we do see a crude design in the events around us and in our own lives. But should we not stop our analysis there? Can we not simply say that the world makes sense to us and that we can operate safely within its rhythms? From empirical data it would appear that we could hypothesize the existence of a crude design or declare that a sense of reason pervades our experiences. But any further statement would be speculative and should be designated as such.

Here evolutionists are much closer to the proper interpretation of the data than are theologians, for they claim that we can at best observe processes in today's world. We cannot, with any degree of certainty, hypothesize or discern a purpose in the cosmic process other than to observe that it appears to be running down. Yet to deny the existence of a creator, let alone a divinity, seems like madness within the Western religious realm. Even some American Indians today would be shocked at such a proposition. Many Indians no longer speak their own languages, however, and most tribes have been under the influence of Christian missionaries and educational systems for nearly a century and a half, so they have entered the conceptual world of the West and now pray to a creator. Had they not been subjected to Western influence, they would probably be using the word or set of words in their own language that designates the great mysterious energy found in nature. Those terms would be empirically derived from unusual spiritual experiences. They would not be the product of a logical thought process.

It is sufficient to note that the monolith of "religion" that is said to confront science as a competitor is considerably less than a unified body of knowledge. Most concepts used to describe the traditional ways of other peoples are applicable only in a limited number of interpretations. When we examine what the various traditions have in common, we find that the non-Western religions' emphasis on nature brings them considerably closer to modern scientific insights and definitions than to Christian concepts. Not suffering the bifurcation of knowledge, their concepts reflect a simple desire to understand the world without filtering their knowledge through the complex intellectual apparatus of the Western tradition that sets religion apart from and against science.

The quarrel between evolutionists and creationists focuses on the explanation of a possible Earth history. Was it long and tedious, featuring gradual or even rapid episodes of organic growth from tiny molecules? Or was it a sudden creative blossoming of life forms with or without a creator? We have projected a six-billion-year expanse of time in which evolutionary changes occurred. But these changes are so minuscule that they cannot be detected. With evolution we do not have a history but merely a hopeful belief. The flaw in both Western scientific and religious thinking begins with the reception of the Old Testament by early gentile Christian converts in the Greco-Roman world. Accepting Genesis as the exclusive explanation of planetary history, they embraced the idea of a linear unfolding of cosmic time beginning in the garden of Eden. St. Augustine firmly implanted the idea of the absolute progression of time in the Western mind so that it became a philosophical constant. Science simply appropriated linear history from Christianity when it sought to answer the question of origins. That appropriation now forces us to link everything in one grand temporal scenario in which life struggles from single-celled creatures to the complexity we find today.

Many of the other religions tell us of occasional or periodic destruction of the planet and the subsequent revitalization of life. The Jewish scriptures accept a multiplicity of worlds. There is little doubt that we can observe a succession of world eras in the geological strata in which radically different conditions and biotic systems existed. Linear time can probably be applied within these eras in a projected chronology. And the eras generally form a sequence so that a crude Earth history can be constructed. But can we say with any justification that there was sufficient continuity between the eras to ensure the progressive evolution of organic life? The "sudden" emergence of fossils in strata does not suggest continuity. Don't we delude ourselves when we posit long stretches of time existing within and between these geologic eras? The truth is that we simply do not know.

What distinguishes Christianity, Judaism, and Islam—the people of the Old Testament—from the other traditions is that, as Ian Barbour suggests: "In the Western religions, myth is indeed tied primarily to historical events rather than to phenomena in nature."[233] Myth here must be regarded as a fictional account or an educated guess based on the best evidence generated by a particular society or culture. But can we have "historical" events that do not have physical, planetary significance? We must insist on this qualification lest we wander into the thickets of "Just So Stories" where everything again is a matter of personal choice. So while Noah's flood may qualify as such a planetary event, many other incidents in the Old Testament would not be included unless they were linked to significant physical, planetary happenings.

In his book *Theology for the Third Millennium,* Hans Küng suggests that the coming theology must be one where "human history is to be synchronized with the history of nature, in order to arrive at a new viable symbiosis between human society and the natural environment."[234] I suspect it is this desire that has prompted Ian Barbour, Paul Davies, Keith Ward, and many other thinkers to try to link contemporary science (a uniformitarian process but not a history) with the Christian interpretation of history (a highly selective series of events affecting a minuscule number of people). And indeed, we have already seen instances in which they are content to accept scientific doctrines uncritically and attach a short Christian view of human history, pretending they have synchronized the two.

History, however, is a treacherous thing to handle. When we celebrated the year 2000 of the Christian era, the American news media seemed to believe that the date was something portentous. Actually, we cannot conceive of what the "real" date might be or how we might define where to begin our dating. The Chinese, Indians, Jews, and others have much older calendars than do Europeans. If we took human history seriously, we would begin not with Genesis but rather with the earliest possible date we have for human habitation and attempt to fill in the timeline with specific events. This task is, of course, exactly what the anthropologists claim to do. We face the same question we did with regard to endless cosmic time: What was god doing until the Hebrews arrived? History does not do a whole lot for religion when we are faced with the task of synchronizing science and religion.

Arnold Toynbee gave a classic expression to the problem of Western history: "An equation of Hellenic and Western history with History itself— 'ancient and modern'—if you like—is mere parochialism and impertinence. It is as though a geographer were to produce a book entitled *World Geography* which proved on inspection to be all about the Mediterranean Basin and Europe."[235] Western thinkers might well reply that it has been the West that has developed science, split the atom, and provided startling new developments in medical care. Indeed, Pierre Teilhard de Chardin lists these achievements as part of his apology for science.[236] But the benefits of these technical achievements are actually available to a very small portion of our species. The majority of people on the planet still live reasonably primitive and restricted lives. With the commerce in Western arms and

munitions, they now live subject to dangers they could never have anticipated a century ago. The condition of the world today is hardly encouraging considering the various signs of ecological destruction we are now able
to document. The claim of progress by a few fortunate millions in the
Western industrial nations may also be seen as a claim to blame.

As long as the religious segment of the American population insists that
it alone understands history, there will be conflict between science and religion in America and continuing competition between the Western religions and the other traditions. Judging simply by the secular history that
we have available to us, the Western religions should not be citing history
as the justification for their existence. They do not appeal to the historical
facts as we know them but devise a sacred history that seems to have few if
any real-world reference points. Hans Küng, in *Theology for the Third
Millennium,* emphasizes the theme of theology giving life to historical
events through its special means of interpretation. Let us examine Küng's
effort in this respect. Küng advances some criteria for judging the various
religious traditions. His standard is a *negative criterion:* "Insofar as a religion *spreads inhumanity,* insofar as its teachings on faith and morals, its
rites and institutions hinder human beings in their human identity, mean-
ingfulness, and valuableness, insofar as it *helps to make them fail to achieve*
a meaningful and fruitful existence, "it is a false and bad religion."[237]
Having established this criterion, Küng then cites with approval the statement of Thomas and Gertrude Sartory, Austrian Catholic theologians: "No
religion in the world (not a single one in the history of humanity) has on
its conscience so many millions of people who thought differently, believed
differently. Christianity is the most murderous religion there ever has
been."[238] Presumably this high status is awarded in spite of the blood
sacrifices and genocidal tendencies of other religions. Real, verifiable
history, then, is not the ally of Christianity.

In summary, then, "religion" competes with science because both activities have been separated out from human experience and boundaries have
been created to isolate them from each other. If we look at the vast majority of human traditions on the planet, we find that they represent an
entirely different approach to understanding the unique experiences we
have previously classified as religious. The concern of these traditions is to
find a path or way whereby we can live in the phenomenal, physical world
in the best manner. Today that path or way could as well be derived from
science as from religion if we admit that empirical experiences tell us
something about the nature of the world in which we live. But how can
these sentiments be expressed in Middle American institutions?

CHAPTER SEVEN

THE PHILOSOPHY/SCIENCE OF OTHER "RELIGIONS"

IN MOST OTHER CULTURES of the world, religion does not exist in the form in which we encounter it in the West. Many of these cultures, instead of concerning themselves with obedience to a personal god and worrying about some possible future salvation, seek to find ways of living within the cosmos in the most harmonious manner. Their view of the world is reflected in their architecture, gardening and other agricultural practices, arts and crafts, sculpture, music, and devotional exercises. Instead of remembering historical events of revelation to which can be attached the vision of creation, falling into sin, atonement, and eventual salvation from the physical world and the creation of a new world, these cultures seek an understanding of cosmic process and develop ways to enhance their lives through participation in cosmic rhythms. Is it possible, then, to find in the other traditions that we have formerly designated as religions some ideas and philosophies that would be compatible with the scientific enterprise?

Western thinkers have traditionally interpreted non-Western religions as inadequate understandings of ultimate reality because they have lacked the cosmic courtroom scenario of the West. The practice has been to compare some of the beliefs of other religions with concepts valid only within the structure of Christianity and to thereby demonstrate the superiority of Christianity. Investigation of the religious propensities of humans, therefore, has been not a neutral exercise but a form of apologetics for Western thinkers. To properly understand the nature of non-Western religions, we must develop a means to judge them on the basis of what they seek to accomplish. Since we know that religion does not appear as a viable category in non-Western thinking, it seems logical to conclude that when we turn to other traditions we are confronting philosophies and ways of life rather than the religious categories we find familiar.

Misunderstandings of major proportions occur even when there is a sincere effort to compare religious beliefs across cultural lines. In this chapter we will first examine some Western thinkers' interpretations of the words and concepts derived from other traditions to show that "religion" is not an accurate descriptive term to use to understand other cultures' beliefs and practices. We will then examine these beliefs and practices to see whether they would raise the same kinds of objections that concerned scientists in the creation/evolution debate. This discussion will help us see that the quarrel over origins in America is but a parochial disagreement within the Western worldview.

To illustrate how Western thinkers distort the question of religion and reach dreadful misunderstandings, we can examine Paul Tillich's comparison of Christianity and Buddhism. Stating that both place a negative value on our existence, Tillich argues that "the Kingdom of God stands against the kingdoms of this world, namely the demonic power-structures which rule in history and personal life; Nirvana stands against the world of seeming reality as the true reality from which the individual things come and to which they are destined to return."[239] If we examine the parallel that Tillich has drawn, we find ourselves comparing apples and oranges.

The "kingdom of god" is a term loosely used to describe an idealistic future social state. It is used to judge historical and existing social institutions against an abstract, perfect, and eternal standard. Nirvana, on the other hand, speaks directly to the apparent materiality of the physical world and its processes. Nirvana, or a concept like it, can be derived from the observation of natural processes and the emotional experiences of spiritual personalities. It is an empirical concept with empirical references. The kingdom of god, on the other hand, invokes a political image of social organization derived from previous human experiences and is not a universal concept. Many societies did not have kingdoms, so the image is wholly foreign, especially to small tribal societies. Tillich finds a sense of the negative in the two traditions and believes they are trying to answer the same question. He then equates the answer each culture provides.

Reading Tillich's comparison from our usual perspective, we would typically condemn the Buddhists not just for adopting a negative view of life but for rejecting the reality of the physical world. The Buddhists, however, are not describing "religion" as such, or even analyzing the human condition. They are commenting on the reality of the physical world—a task that Westerners leave to their physicists. The Buddhist statements could therefore qualify within the Western understanding as either religion or natural science. How close would nirvana be to the theories of physics? Rupert Sheldrake reports, "Einstein said that the entire universe consists of a gigantic universal field, the gravitational field. This field contains everything that is within the universe: it interrelates everything in the universe. And the gravitational field is prior to matter."[240] Do not nirvana and Einstein's gravity field have a close relationship? Is this definition what the Buddhists, using more technological, scientific terms, would describe as nirvana? If matter is a product of this field and ultimately illusory for

Western physicists, should the Buddhist idea even be put into a religious category? And why do Western theologians insist that the physical world is ultimately material?

Discussing the Hindus, Tillich declared: "In India the Brahman experience and speculation deprive all things in time and space, gods as well as men and animals, of their ultimate reality and meaning. They have reality—*but from the point of view of Maya;* they are not simply the products of imagination, but they become transparent for the ascetics who have discovered the principle of Brahma-Atman in themselves and in their world."[241] (Emphasis added.) Again, there is no discrepancy here. Adopting the point of view of maya places the Indians in a position to speak to ultimate questions, but the conclusion they draw is not necessarily religious, as we understand it. Do not our physicists draw all kinds of conclusions about the physical world from many points of view? Affirming the material reality of gods, men, and animals does not make them ultimately real in view of evidence to the contrary. That insight, that there is no ultimate status for entities at a certain level of existence, could also be understood as a scientific statement rather than a religious belief.

Ian Barbour stresses the same point as Tillich when discussing the Eastern traditions. He says that "especially for the Advaita tradition in Buddhism, the temporal world is illusory and ultimate reality is timeless. Beneath the surface flux of *maya* (illusion) is the unchanging center, which alone is truly real, even though the world exhibits regular patterns to which a qualified reality can be ascribed."[242] Is this explanation positive or negative? How is this belief significantly different from that of modern physicists analyzing the material world? We usually read the description as negative in light of our unexamined belief that the material world has some ultimate reality and believe that it is foolish for Buddhists to say it does not. Is it necessary to affirm the ultimate reality of matter to qualify as a religion in Tillich's eyes?

Statements from non-Western cultures can be positive or negative depending on how Western minds classify them in the Western scheme of things. If we took the so-called religious statements of the East and spoke only to the Western scientists, the statements would have status. When we classify them as religious, they shrink in value considerably. David Foster writes, "When one looks for the fine-structure of molecules as to their atoms, one enters a region dominated by void or emptiness; and ultimate fine-structure is not structure at all but consists of electrical and gravitational fields cavorting in the void."[243] Unquestionably we are describing maya and nirvana. One might well classify Heisenberg, Bohm, Einstein, and Bohr as Western mystics, or even as devout Hindus, who refuse to deal with the "real" world and insist on declaring the physical world an illusion.

There is such schizophrenia in the Western mind regarding science and religion that very bright thinkers distort ideas outrageously. On the witness stand at the Arkansas trial, Langdon Gilkey explained for the court the difference in creation theories between Judeo-Christian beliefs and those of other religions. Only the three Western religions, he said, talked about creation out of nothing. Other traditions, he explained, described their origins in significantly different terms: "Our world emanates from the divine as an appearance of God; our world is *maya*, an illusion, a dream; our world arises out of two independent and equally ultimate principles: matter and form, one divine and the other not."[244] It seems that Gilkey is describing, in reasonably accurate terms, English physicist David Bohm's "implicate order," where mind and matter are two manifestations of a deeper reality.

It is a pity that no one in the courtroom was alert, able to see the similarity, and inspired to inform the court of the identification. What would the response have been by the evolutionists and creationists to such a proposition? Would the judges have had to prohibit the teaching of quantum theory on the grounds that it represented an Eastern religious doctrine? It is not difficult to see that we are approaching a critical time in human history in our search for unifying principles. The fragmentation of knowledge into separate disciplines must be repaired, and the insights of many cultures must be granted admission to the intellectual discussion.

The mind/matter equation that suggests the existence of an implicate order might be better expressed by contrasting mind and energy (a form of matter, according to physicists). Energy is more suitable for our discussion since matter is such a commonplace phenomenon that it does not invoke any feelings that might be classified as religious. Energy, however, is another thing altogether. Today we talk about "fields" of energy that compose the universe, and if matter is actually energy dancing in a matrix that can be described only by mathematical formulas, we need no longer oppose mind and matter as if both had some ultimate reality.

The framework adopted by Western thinkers a century ago for interpreting religion was the cultural evolutionary incline that our species was believed destined to climb, creating tools and language and becoming capable of increasingly complex thought. At the bottom were the primitives who were able to intuit a pervasive energy that made things move. These people were thought to live in constant fear of natural forces that exhibited unrestrained energy. Then spirits were invented, eventually they were transformed into gods, and finally the experience of mysterious energy was encapsulated in the modern Western notion of one god. The ability to discuss this god in abstract terms came to be regarded as the ultimate knowledge of this mysterious energy. Thus formal theology reached its high tide with Tillich's "god above god."

Many scholars uncritically assumed that at each stage of development each new apprehension of this energy was a product of increasingly sophisticated intellectual activity and that our species was evolving in its ability for abstract thought. Keith Ward illustrates how Western thinkers have used this cultural evolutionary framework to describe other religions. He says that the Eastern religious stream "develops from forms of animism to the idea of a cosmic order, a way of balance and harmony, following which brings stability and calm of mind, and peace and right order in society."[245] If the first apprehension of higher powers occurs in animism, which would be the recognition of the mysterious energy of the universe, it is not such a far leap forward to then intuit an orderly arrangement of the energy that will bring balance with the rest of the cosmic process. This can be understood as a fully developed natural theology, although it does not have a place for a god. It might also describe a nonmechanical way of understanding the energy that would be complementary to physics. The critical idea here, though, is that animism is supposed to represent primitive superstition when we have no evidence that it does.

Paul Tillich spends considerable time outlining the features of animism to make certain that the primitive label is affixed firmly. "The conception of nature that we find earliest in history, so far as we have knowledge of it, is the magical-sacramental conception," he writes. "According to it, everything is filled with a sort of material energy which gives to things and to parts of things, even to the body and the parts of the body, a sacral power. … At this phase of cultural development the distinction between sacred and profane is not a fundamental one. The natural power in things is, at the same time, their sacral power, and any commerce with them is always both ritualistic and utilitarian."[246] Linking "magical" with the sacramental is objectionable here because of the pejorative connotations, and it represents a misunderstanding of rituals and ceremonies on Tillich's part. Instead of calling the feats of shamans, yogis, and others "magic," we would do well to contemplate how these people can perform feats that appear to deny our cause-and-effect expectations. If the early religious view is sacramental, it is an attitude that requires a specific approach to the world and its beings. It recognizes that they have value in and of themselves.

Van Der Leeuw, in his study *Religion in Essence and Manifestation*, offers an insightful corrective to Tillich's ill-advised effort to describe the beginnings of religion as primitive magicalism. "What we moderns call 'nature,'" he explains, "has a prominent role in all religions without exception. Yet it is neither Nature, nor natural phenomena as such, that are ever worshipped, but always the Power within or behind."[247] And Van Der Leeuw further explains that for the so-called primitive peoples, who are identified as exemplifying the earliest evolutionary scenario, "the Power in the Universe was almost invariably an impersonal Power. Thus we may speak of dynamism—of the interpretation of the Universe in terms of Power."[248] The simple substitution of "energy" for "power" in these sentences would place the so-called primitives in the same camp as the modern scientists. For both, the apprehension of power would be an experience from which an insightful conclusion could be drawn. Any specific and consistent way of gathering data and using it to enhance one's ability to understand the world should be seen as compatible with the scientific method.

Until Christianity changed the focus from nature to history, the idea of power was accepted by a substantial percentage of the world's human societies. It was called many things but always referred to the underlying power of the universe. In many sociological and anthropological textbooks we encounter the concept of "mana," which is presented in a pejorative sense as an early superstition. But across the spectrum of human traditions outside the West, religious people have experienced or intuited the same energy of the universe. H. R. Hays, popular prehistorian, discussing the Romans, says, "The multiplicity of gods resulted from a habit of mind which discovered *numen* in practically everything. The concept of *nimuna* approaches very closely to that of mana, the electrical god stuff that could be identified with a symbol, an instrument, an animal or a man."[249] And he states that the *Silua* or *sila* of the Eskimos "is vague and remote but corresponds pretty close to mana."[250] F. David Peat writes, "The idea of a High God appears to have little significance to the Naskapi. Rather it is Manitu which assumes importance. Manitu is the essence of all things and resides in them."[251] Even the Aztecs have a term, *teotl*, which "seems to apply to the supernatural in general and has pretty much the significance of mana."[252]

If we were to conduct an exhaustive survey of popular books written by scientists trying to explain the quantum world in which some mention of or allusion to the similarity of the primitive mana and the energy of modern physics occurs, we should have a long list. Fred Alan Wolf summarizes the connections between the two conceptions for the popular science writers of the future: "Today, our position is close to the one discovered by basic tribal peoples. The concept of universal energy in our language might be called the 'universal quantum wave function' or 'matter wave' or 'probability wave' of quantum physics. This 'wave' pervades everything and like the universal energy, it resists objective discovery. It appears as a guiding influence in all that we observe. *Perhaps it is the same thing as* the 'clear light'—the all pervading consciousness without an object—of Buddhist thought."[253] The energy may indeed be mind/energy if it is a guiding influence.

The complaint may be made that so-called primitive peoples could not have described energy in the terms used by physics. But if, based on their experiences in nature, they were able to discern the ultimate constituent of the universe without recourse to sophisticated instruments and complicated mathematical formulas, that only demonstrates that these peoples' languages and experiences gave them a deep knowledge about the world. What would be the effect if this connection between primitive man's ideas and quantum theories were taught in Arkansas and Louisiana schoolrooms? Would it violate the separation of church and state? If presented as religious belief it would cause trouble, but if presented as physics it wouldn't. The problem is in our way of thinking about things.

Animism is supposed to generate many gods and lead to polytheism. But it is difficult to find such progressions in human history. Shintoism, in particular, creates problems for Western thinkers in this respect. Ernest Benz noted that the evolution toward monotheism was far away from people in that tradition: "Its 800,000 gods have hardly been put into a hierarchical order, each god being a particular manifestation of the Numinous itself."[254] Surely the concept of god is useless when gods are this plentiful. It seems likely that the scholars who translated Shinto beliefs into European languages mistakenly used the term "gods" when they should have said "spirits" or "energetic entities." Shinto's multitude of spiritual personalities more probably represents the memory of occasions when the cosmic power manifested itself in concrete events or in the lives of outstanding spiritual people than an incredible pantheon of competing deities.

Nathan Söderblom pointed out that ancient Greece, the birthplace of Western philosophical thought, never evolved to monotheism. In fact, he cited with approval Italian history of religions scholar Raffaele Pettazzoni's argument that monotheism is "far more a question of some particular event, which has occurred but a very few times in the history of the world (the great majority of the peoples have not become monotheistic by evolution, but by conversion, i.e. when they embraced a new and foreign religion which was monotheistic) and this was always connected with the life and work of a great religious personality."[255] Even within the Christian tradition we have examples of forced conversions. Moses received the law at Mount Sinai while below the Hebrews made a golden calf to worship. He then ordered the slaughter of golden calf worshippers, illustrating Pettazzoni and Söderblom's contention, and leaving the monotheists in charge. Perhaps this is an extreme example of punctuated religious evolution? We must rid ourselves of the belief that polytheism automatically evolves into monotheism or that monotheism is anything more than a doctrinal political statement seeking to unify societies.

It seems very unlikely that animism evolves into polytheism since we have no known instances of it historically. Van Der Leeuw makes an important point when talking about the universal Power discerned by many societies. "The power which, nameless, moves within the Universe," he counsels, "is ultimately *One*; that is, there is none other beside it. Here— not in Monotheism—unity receives its full stress."[256] He elaborates on this phenomenon, observing, "A certain Monism already constantly present but concealed by practically oriented primitive thought, now rises unmistakably into view; and what has hitherto been erroneously maintained about the actual idea of Power becomes quite correct—namely that 'this interesting sketch of a unified apprehension of Nature and of the Universe reminds us, in virtue of its principle of unity of Monotheism, and in the light of its realism, of dynamic Monism.'"[257] If we find a fundamental monism in the apprehension of power, why would we move tediously to polytheism and then seek to impose monotheism on ourselves? We must carefully distinguish between monotheism, which speaks primarily of the nature of deity, and monism, which describes the experience of encountering the underlying energy of the world.

H. R. Hays, in his book *In the Beginning*, develops this idea in a different fashion and provides us with the insight to understand the nuances here. "There is no indication of true monotheism, the worship of a single, exclusive, jealous god, anywhere among the very early peoples," he writes, "or peoples still in an early stage of culture."[258] Monotheism is an intrusion into an otherwise coherent adjustment to the universe. Recognizing the difference between the two ideas, we see that monism is compatible with modern science; monotheism must be inserted in a scientific scenario by force or persuasion. Modern apologists who seek to unite science and religion would do well to start with other traditions and, after reconciling them to modern science, find a place for Christianity afterward.

Paul Davies recognizes the dilemma of modern Christian apologists. If they hew too closely to the concepts of modern physics, they tend to support the old concepts of animism and monism as adequate descriptions of the world. If they do not speak to the basic concepts of physics, their theology becomes an abstraction unrelated to cosmic processes. In *The Mind of God*, Davies attempts to solve this problem by resurrecting the old pantheism in a new word to distinguish non-Western traditions from the

West. Pantheism, in his estimation, means that "god is identified with nature itself; everything is part of God, and God is everything."[259] This word is supposed to represent the concept of non-Western peoples. He offers "panentheism, which resembles pantheism in that the universe is part of God, but in which it is not all of God. One metaphor is that of the universe as God's body."[260] This concept is too transparent. Conceiving the universe as god's body moves far away from all religious traditions, most of which did not take speculation this far.

Ian Barbour suggests that modern theology has come to grips with the changes in our perception of the world revealed by science. He believes that "the evolutionary view of nature molded the modernist understanding of God; the divine was now an immanent force at work within the process, an indwelling spirit manifest in the creative advance of life to ever higher levels."[261] This definition has problems in that evolutionists eschew the idea of purpose in the development of cosmic and organic life and rely instead on chance. To have an "indwelling spirit" working purposefully violates the basic assumptions of science. The biggest problem for apologists such as Ian Barbour, Keith Ward, and Paul Davies is that they endorse the discoveries of modern science (primarily physics) as far as they can, but they then insert the concept of the Christian god. Surrender the god concept and the problem is solved: We simply return to the primitive tribal/modern quantum statement that the basis of this apparent physical universe is a mysterious energy that manifests itself in mind and matter.

Western thinkers have generally seen worship as the primary relationship of humans with gods. From animism to polytheism to modern monotheism, religious activities of non-Western peoples were regarded as forms of worship. But the so-called primitives did not worship in the sense that we understand the word today. Rather, they petitioned the higher powers to take pity on them and assist them. Eastern devotional activities were more often the ways and practices that would enable people to connect with cosmic processes. If we deny the Western religious concept of god, then how do we explain the activities of worship that we have traditionally associated with religion?

From the earliest times non-Western peoples have engaged in specific kinds of rituals to establish relationships with higher powers. Indeed, Robert Bellah notes that "primitive *religious action* is characterized not ... by worship, nor, as we shall see, by sacrifice, but by identification, 'participation,' acting out."[262] Sacrifice and worship may be minor parts of the larger sacred activities, but they are not central to religious belief. This difference is fundamental for understanding how non-Western peoples have not required the Western concept of god. Participation enables the human and the higher powers to act jointly and cooperatively in creating sacred events. Once a medicine man or shaman has received powers from a spirit, a covenant is established and he is thereafter able to heal, predict the future, and demonstrate the nature of his power empirically.

The Christian practice of formal worship, when viewed in an objective light, looks very strange. Hymns of gross and transparent flattery abound, in which the worshippers inform the deity that he is the best possible and most powerful god—in a sense reassuring him that he is god. Historically, the Western god has been almost pathologically jealous of other gods, and continual praise is necessary lest his anger break loose and cause great harm to the world. It goes without saying that the deity already knows about his talents and virtues, so except for disciplining and controlling his followers, the adoration is unnecessary.

The personality of the Western deity is difficult to visualize if he continually and unabashedly requires lavish attention. People are so accustomed to praising him that they never ask why they feel obligated to relate to the deity in this manner. It could be merely appreciation and thanksgiving, since the deity in America is expected to make us rich and slender, to intervene in Texas high school football games, and to endorse whatever path of action our government may wish to take. When viewed from the perspective of other traditions, the Christian form of worship seems crude and at times blasphemous. Terms of grandeur are often used in other traditions, but their goal is establishing a balance in people's lives through cooperation with higher spiritual forces. Good and bad do not enter into the equation as a rule.

Worship may well be a result of historical events. In the early stories in Genesis we find people negotiating with the deity and standing in some sense as equals before him. Yet there is a change during the course of Jewish history such that in Isaiah's vision, at least a millennium later, we see god as an oriental despot in his palace. At least some of the structure of the Western religions must come from some unusual incidents in Earth history, the nature and intensity of which we have forgotten. Robert Bellah calls our attention to "the emergence in the first millennium B.C. all across the Old World, at least in centers of high culture, of the phenomenon of religious rejection of the world characterized by an extremely negative evaluation of man and society and the exaltation of another realm of reality as alone true and infinitely valuable."[263] And he points out an "equally striking fact, namely the virtual absence of world rejection in primitive religions, in religion prior to the first millennium B.C. and in the modern world."[264] Here is a clear change in religious perspectives within historical times that we can examine.

The phenomenon of adopting a negative view of the world did not affect the primitive peoples, who certainly must have represented the majority of human beings in the world at that time. It seems to have affected those societies that had adopted and developed an urban cultural/economic way of life. We should therefore look for some historical event in the equatorial latitudes so traumatic that it convinced civilized peoples that the physical world was unreliable. Perhaps a major physical catastrophe was responsible for this change of viewpoint. Profound political changes might have occurred. Worship of gods may have originated as a result of this event.

Even more remarkable is another period centering around 550–500 B.C. that seems to have been equally important in the development of world religions and philosophies. Nathan Söderblom, describing this period, noted: "Kon-fu-ste summarized the wisdom of China in reverence towards Heaven and antiquity.... Lao-tse showed the way to quiet goodness and peace of mind. On the banks of the Neranjara there rose up in the soul of Siddhartha the way of deliverance from the woe of existence and the attainment of Nirvana.... From other mysterious sources came salvation by Bhakti.... Among the Jews appeared the great prophets and their writings.... Heraclitus expounded his doctrines.... Pythagoras gathered an intimate band of hearers to hear his speculations.... Xenophanes spent the greater part of his life.... In Sicily the seer Empedocles proclaimed his own divinity...."[265] Obviously this time period produced the philosophical frameworks that characterize the formal teachings and practices of non-Western civilized religions and that eventually produced secular philosophy.

What happened to motivate people in so many societies to seek ultimate answers about the world? Does the Jungian collective unconscious apply here? If we were thinking about the nature of the cosmos, worship might have been the common peoples' practical response to a crisis in seeking a better understanding of life. Why was there a need to express apprehensions about the cosmos in an abstract fashion? The study of religions should perhaps begin with the religious innovations of this period rather than with a mythical story of cavemen cringing when they saw lightning, for which we have no evidence. Why did not the tribal peoples also take a leap toward expressing their experiences in abstract terms? Perhaps it was not possible to attach their concerns to a formal political structure in order to find stability.

Some of the traditions arising from this fertile historical period are interesting in that they provide useful bodies of knowledge that can easily speak to modern science. Buddhism, for example, is compatible with quantum physics as long as it is not seen as wholly devotional. The Confucian philosophy and Shinto offer easily understandable ways of dealing with the world. Where the Buddha sought to understand the cosmos and our place in it, Confucius concentrated on the world of human activities and sought social and political reforms. If we place his teachings in a Western context, Confucius was concerned about the *oecumene*, the world of people, and he sought to find the key to bringing peace and harmony among individuals and institutions. The greater cosmology of the Chinese probably had much similarity to the Indian teachings, but it was not important to him. In this sense his philosophy was much like Christianity. According to Huston Smith, renowned historian of religion, Confucius was distressed that "men had become individuals, self-conscious and reflective. This being so, spontaneous tradition—a tradition that had emerged unconsciously out of the trial and error of innumerable generations and held its power because men felt completely identified with the tribe—could not be expected to command their assent."[266] The answer was to establish proper authority by emphasizing and clarifying the ancient traditions that had previously held societies together with some modicum of justice.

Confucian philosophy and its tenets seem reasonable, and for Americans, with their alleged conservative beliefs, they should be very popular. Strangely, the separation of the individual from tribes and communities has been seen as a positive value, indeed the essence of Christianity, for some important Western thinkers. Robert Bellah, for instance, says: "The identity diffusion characteristic of both primitive and archaic religions is radically challenged by the historic religious symbolization, which leads for the first time to a clearly structured conception of the self. Devaluation of the empirical world and the empirical self highlights the conception of a responsible self, a core self, or a true self, deeper than the flux of everyday experience, facing a reality over against itself, a reality which has a consistency belied by the fluctuations of mere sensory impressions."[267] Here we have the emergence of the idea of an independent self but at the cost of an isolation so severe as to create societies in which personal responsibility becomes a burden too heavy to bear and individuals attach themselves willingly to institutions that then direct their lives.

Paul Tillich would have enthusiastically endorsed Bellah's analysis. In *The Protestant Era*, after describing primitives as having a sacral power that binds together family, rank, tribe, neighborhood, and ritual community, he describes the individual as being "swallowed up" in the all-embracing unity of the group. He lauds the destruction of this communal cohesion: "All this changes when the system of powers is replaced by the correlation of self and world, of subjectivity and objectivity. Man becomes an epistemological, legal and moral center, and things become objects of his knowledge, his work, and his use." [268] Christianity is said to appeal most to the individual who breaks from family and tribe and enters a larger and less homogenous ecclesiastical community. This change, however, is simply one of rendering a living universe into a lifeless gathering of things to be used by humans indiscriminately. It does not correspond to what people feel emotionally.

The sad thing about Christian apologists is that they must defend their faith at all costs and find ways to glorify it. Since it does not logically hold together, they often make conflicting statements that betray their arguments. Let us return for a moment to Tillich's praise of man as an epistemological, moral, and legal center. It isn't as noteworthy as he leads us to believe. "Man who transforms the world into a universal machine serving his purpose," Tillich writes, "has to adapt himself to the laws of the machine. The mechanized world of things draws man into itself and makes him a cog, driven by the mechanical necessities of the world. The personality that deprives nature of its power in order to elevate itself above it becomes a powerless part of its own creation."[269] True—and thus the concern of Confucius was valid.

Few Western thinkers have bothered to examine the Shinto tradition closely. Perhaps its relatively late appearance as an identifiable way of life (around the fourth century A.D.) or its alleged borrowing of concepts from Confucianism and Buddhism make it suspect. Nathan Söderblom, Huston Smith, and Rudolf Otto, German theologian and mystic, apparently felt that Shinto was so radically different from the other religions that it should not properly be included in the same category. It has always seemed to me that Shinto, with its concern for beauty and landscape and its balance between the divine and the natural, represents a process of development that a North American Indian tribe would have experienced if it had grown immensely in population and needed to formalize its customs and traditions in a more secular fashion.

According to Robert Bellah, the general background beliefs of the many sects that are identified in some way with Shinto are very much like the views of the so-called primitive peoples. "Man is the humble recipient of endless blessings from divinity, nature, his superiors, and quite helpless without these blessings. He is a microcosm of which divinity and nature are the macrocosms."[270] At its core, then, Shinto is ritualistic with a minimal development of dogma. It can speak to the natural world in ways that other traditions cannot. As such it would appear to be compatible with either evolution or creation but also might refuse to provide any answer to the question of origins.

The Eastern religions, then, do seem incomplete if viewed in the traditional Western format of emphasizing belief in dogma, of elevating a local history to be the sole example of the passage of divine cosmic time, and of developing a rigid institutional setting in which religious activities must take place. It seems clear, however, that they are better attuned to dealing with the real questions of cosmic existence because they confront the whole range of human experience and knowledge. "Salvation," if we can call it that, in these traditions is directed toward playing one's designated part in the movement of the cosmic process. In the West, salvation looks to the destruction of the physical universe and the creation of a new heaven and Earth. But there is no guarantee that this new creation will be better than the original one, which had to be destroyed by water because it had become wicked. The only guarantee is that a big flood won't destroy us.

When I was in seminary three decades ago, a newspaper poll, one of those meaningless things taken by Americans to determine truth, announced that the two most Christian lives in the twentieth century were those of Albert Schweitzer and Mahatma Gandhi. We had a fine time tweaking the noses of the more fundamentalist students about the idea that the two most Christ-like lives were those of a nonbeliever and an apostate Christian who had produced an embarrassing but highly realistic portrait of the historical Jesus. Today, almost unanimously, Christian theologians seem to have adopted the stance that moral and ethical behavior equal and sometimes superior to Christian behavior can be found in any of the other traditions. Thus James Gustafson, writing an introduction to Protestant theologian H. Richard Niebuhr's *The Responsible Self*, admits: "The Christian community cannot claim a superiority for its moral knowledge, or for its capacities to cultivate moral wisdom, or for its power to determine for others what the proper course of their affairs ought to be."[271] Ian Barbour, John Macquarrie, and

H. Richard Niebuhr all agree wholeheartedly, and Gustafson even begins his own book with a description of an incident in which a nonbeliever acted with more moral sense and compassion than he, a devout Christian, did.[272] So one cannot argue convincingly that teaching creation would create citizens who are more responsible or who have a greater appreciation of the world, nor that other traditions have a lesser ethical sensitivity than Christianity.

Ian Barbour offers a caveat to this general consensus, however, with an unfortunate, gratuitous statement that cannot be proven empirically. "Natural theology," Barbour says, and surely we are talking about natural theology when we speak of the beliefs of the other traditions, "does not lead to the personal involvement and dependence on revelatory events which characterize the Western religious traditions."[273] This comparison is badly misconceived. In almost every other tradition we see personal commitments to the higher powers far in excess of the commitments of almost all Christians. There are no vision quests, very few fasts, a minimal number of mystics, and sparse efforts to engage in devotional practices within Christianity. At least in the United States, it is difficult to distinguish the Christian religion from secular society. Revelatory events occur continually in the ceremonies of other traditions, although they generally deal with practical problems and do not predict the end of the world.

After this brief review of the other traditions and a comparison of some of their beliefs and practices with those of the Western religions, it should be apparent that the idea of "religion" is badly conceived and should be restricted in its use to those countries in the West who wish to maintain the artificial division between science and religion. Most beliefs credited to non-Western traditions as religious are actually their form of philosophy. The cosmos they describe has much in common with quantum physics. In the popular expression of these philosophies, their followers have developed a devotional and ceremonial life that expresses the cosmic principles they understand. The majority of religious traditions in the world—from small tribal practices to worldwide followings—have no quarrel with modern science and pose no threat to its activities.

As the United States becomes more heterogeneous in its religious constituency, and as we continue to encounter other traditions and their practices, the concept of religion that we use in law and social behavior will become increasingly blurred. We will have great difficulty ensuring that freedom of religion for practitioners of non-Western traditions is respected, as is required by the constitution. We may decide that beliefs must be protected but that practices can be banned. Or we may allow certain practices but refuse to allow the articulation of certain beliefs. It should be apparent that the other traditions have much to contribute to the scientific enterprise and that the conflict with religion that we presently experience is applicable only to the intellectual inconsistencies within the Western paradigm. An interesting test case might be for a teacher to include readings from a Buddhist text when discussing the origin and age of the universe. Would the courts say, if experts brought to their attention the idea, that the Buddhist materials were not really religious in the same way that the Bible is thought to be?

CHAPTER EIGHT

THE NATURE OF HISTORY

THE QUARREL BETWEEN evolutionists and creationists, and between Western and other religious traditions, reduced to its most basic form, involves the interpretation of history. Two views, irreconcilable with each other, vie for our approval. Did the universe begin a short time ago through a divine act, or did it begin seventeen billion years ago with a sudden explosion of energy? Perhaps the universe simply fluctuates, contracting and expanding according to the new string-theory interpretation. Or, if the universe is really a gigantic complex thought, who is thinking it? The Christian fundamentalists believe the accounts of Genesis literally and seek to reconcile everything we know about the natural world with a very short biblical time span. Clearly, we are facing irreconcilable conflict over the history of the universe and all of its subsequent small processes.

The reigning elite of modern science insists that an endless amount of time can produce almost anything. Species are thought to develop through minuscule innovations in their physical structure via mutations in their genetic makeup—or disappear from the strata, quickly and radically evolve, and then return to the stage in later geological strata (punctuation). We arrange geological strata by identifying fossils. We move from simplicity to complexity, recklessly estimating the time that might have existed between compatible sets of strata. Once a scenario is devised that can withstand the critique of esteemed older scholars, a chronology is approved, and we come to believe that we have accurately described the history of our planet. These accepted scenarios, however, need as much faith as does the biblical story.

Insuperable barriers exist that call into question the reality of both versions of Earth history. The Bible tells us only about the historical adventures of a small group of desert tribes who briefly established a small kingdom amid the larger empires. It describes their political conflicts in the Middle East between approximately 2400 B.C. and the beginning of the Christian era. This history is said to reveal a divine plan for all of creation, including humanity, devised prior to Genesis, that will be consummated in the future with the destruction of the physical world and its transformation into a new and strange world, in which the lion will lie down with the lamb, reconciling the prey-predator relationship that we find in the empirical world.

Missing in this biblical scenario is an acknowledgment of the importance of the histories of other peoples and whatever divine plans they believed their experiences had revealed. In secular scholarship this broader version of history is a major part of the human story. We are interested in tracing the history of the large early empires, beginning with the Sumerians and followed by the Assyrians, Babylonians, Persians, and others, in proper archaeological sequence. We then focus on the Greeks, Romans, Europeans, and English, finally arriving at the Americas. There is no question that each empire believed it was particularly blessed by the acknowledged deity of the time as it reached its zenith. Yet the Hebrew version of early Earth history has become dominant thanks to the evangelism of the early Christians. They brought the Old Testament to the gentiles to provide a historical basis for belief in Jesus Christ, which led to the gentile world's uncritical acceptance of the belief that the physical world had been created or formed only once and that cosmic time was linear. With the victory of Christianity over competing religions in the Roman world, Genesis became the official version of Earth history. Peoples of the succeeding European empires understood the Bible as revealing a divine plan of which they were always the contemporary beneficiaries. With the increase of secular knowledge and the separation of religion from science, the idea of linear time was uncritically accepted as the proper framework within which the physical world could be understood—because science was merely offering secular alternatives to sacred concepts. With the triumph of evolution, linear time has become the primary framework of science; we flee to the idea of endless eons of time when we are unable to explain anything in the findings of paleontology, geology, archaeology, or evolutionary biology.

While we do admit that ancient societies possessed calendars that originate much earlier than those of the Hebrews, we arbitrarily reject the value of their accounts, although we are unable to explain why they would devise calendars of great antiquity if they had no memories of remote times on the Earth. Sumerian stories, for example, go quite far back in calendar years, to a time when the "gods" lived on Earth alone.[274] We also have numerous stories of once-existent continents with flourishing civilizations in the middle of the Atlantic and Pacific Oceans that were destroyed by sudden cataclysms. We usually reject the accuracy of such accounts or seek to bring the legends forward by forcing them into an acceptable modern time frame even if, as in the case of Atlantis, we have to change the locations and dates.

Within the current acceptable scientific stories relating the history of our planet, we also have problems. A century and a half ago, when the biblical worldview was breaking down, the prevailing geological paradigm saw Earth history as catastrophic—although the catastrophe was limited to Noah's flood, since it was described in Genesis and was later discovered to be present in the records of many other peoples. Working with the hypothesis that observable, present-day geological processes had always been the same, and that catastrophes had never occurred on any sizable scale, Earth history was understood by science to consist of billions of years during which minuscule changes had occurred. Comparing the annual accumulation of sediment in lakes and river deltas laid down according to uniformitarian processes, scholars began estimating how long it would have taken for certain strata to form. Thus, estimates of the age of some sandstone and limestone formations ran into the millions of years. There was no guarantee that climatic conditions had remained stable for those long stretches of geologic time, but that didn't seem to bother anyone. Much of scientific dating is reminiscent of Mark Twain's famous essay proving that someday the Mississippi would be many thousands of miles long, extending far into the Atlantic.

Eventually, so many strata were studied that larger classifications had to be devised to describe the various phases that geologists believed accounted for all the visible formations. Today the nomenclature is quite complex. Our basic unit is the *eon*. We have four eons: the Hadrean, the Archean, the Proterozoic, and the Phanerozoic. This classification sounds good until we examine it more closely. As it turns out, the Hadrean eon did not exist here on Earth. It is, according to Brian Skinner and Stephen Porter in their textbook *The Dynamic Earth*, "the earliest part of the Earth's history, an interval for which *no rock record is known*. However, *rocks of this age are present on other planets whose earliest crustal rocks have been little modified since they accumulated.*"[275] (Emphasis added.) It seems strange that we could have an "eon" of geologic time with no data whatsoever. The hypothesis that other planets have these rocks is sheer speculative fantasy

unless some geologist can prove he has been there and verified their existence. Support for the reality of the Hadrean raises severe questions about
the empirical nature of geology.

Eons are subdivided into eras that are defined by the life forms found in
the strata. We have the Paleozoic, the Mesozoic, and the Cenozoic. Eras are
then broken down into periods that in turn are divided into epochs. These
classifications are highly speculative since nowhere on Earth is there found
a complete stratigraphic column, and in many areas critically important
formations are missing, with no discontinuities that would explain their
absence. Many geologists will simply admit that the stratigraphic column
is a disaster. Nevertheless, the basic outline of geological timescales does
have sufficient empirical data for it to serve as a framework around which
a new concept of Earth history can be constructed.

The biggest barrier in constructing a new understanding of Earth history lies in our current view of things. We suffer immensely from the sin
of modernism. That is to say, we have adopted the attitude that we know
more about the ancients than we really do. We assume that the fragments
of data that have come down to us represent the high points of the various
ancient civilizations. We judge them while holding the attitude that they
were unable to express the complex theories of today and so constructed
their understanding of the world from superstition. We fail to realize that
the vast majority of important, ancient scholarly works were lost and that
we are basing our judgments on summaries or secondary sources. Our version of Aristotle's philosophy is in fact the summary of notes taken by his
disciples and, with one rare exception, not the writings of his own hand.
Some of the ancient philosophies that we have pieced together are contained only in the writings of later thinkers who quoted them in order to
refute them. Even where we have deciphered clay tablets and bi- and tri-
lingual inscriptions, we are dealing with only a minuscule representation
of what was actually known by ancient peoples.

We forget, to our detriment, that many ancient libraries were destroyed in wars and religious purges, so it is doubtful that we possess even a thousandth of the knowledge that was gathered together, analyzed and edited, subjected to additional commentaries, and regarded as reliable by ancient peoples. Richard Mooney, popular science writer, made up a short but important list of the destructions of the great libraries of the ancient world that should give us a bit of pause: "The library of Pergamus in Asia Minor … contained 200,000 books, all of which were destroyed. When the Romans razed Carthage in the Punic Wars in 146 B.C. they also burned to ashes a library said to contain half a million volumes. The Romans also destroyed under the leadership of Julius Caesar, the Druidic library at Autun, France, containing thousands of scrolls on philosophy, medicine, astronomy, and mathematics. In China the Emperor Tsin Shi Hwang-to ordered all the ancient books destroyed in 213 A.D. Leo Isaurus burned 300,000 books in Byzantium in the eighth century A.D."[276]

These books were not by any means the novels, "how-to" guides, and children's books that make up the bulk of our modern libraries' holdings. They must have represented hundreds if not thousands of years of careful thought, perhaps even some early scientific experiments, and certainly accurate observations by people as dedicated and reliable as the scholars of our own time. One of the great tragedies of our planet's history, the destruction of the great library at Alexandria, deprived us of a library of 700,000 books in scroll form. "The Bruchion contained 400,000 books and the Serapeum 300,000," Mooney noted. "The university [at Alexandria] also included facilities for the study of medicine, mathematics, astronomy, botany and zoology, and it could house 14,000 students."[277] Can we imagine an ancient university with a student body large enough that it could perhaps qualify for one of our major athletic conferences or send a team to the NCAA basketball "Big Dance"? When we consider that ancient peoples devoted considerable time and resources to the accumulation of knowledge, and that education probably consumed a larger percentage of their social wealth than education in America does today, the contemporary Western pride in representing our modern knowledge as a substantial accomplishment begins to fade considerably.

What did these ancient people know? Giorgio de Santillana and Hertha von Dechend describe the contents of some of the astronomical tablets of the ancients in *Hamlet's Mill:* "Now that the documents of the earliest ages of writing are available, one is struck with a wholly unexpected feature. Those first predecessors of ours, instead of indulging their whims with childlike freedom, behave like worried and doubtful commentators: they always try an exegesis of a dimly understood tradition. They move among technical terms whose meaning is half lost to them, they deal with words which appear on their earliest horizon already 'tottering with age' as J. H. Breasted says, words soon to vanish from our ken. Long before poetry can begin, there were generations of strange scholiasts."[278]

They also describe noted Moslem scholar and mathematician Al-Biruni's visit to India a thousand years ago, when he found "that the Indians, by then miserable astronomers, calculated aspects and events by means of stars—and were not able to show him any one star that he asked for." In other words, ancient scholars were struggling with astronomical knowledge so old they had even forgotten its empirical application. And they comment, "The Mayas and Aztecs in the unending calculations seem to have had similar attitudes."[279] Surely we are dealing here with long stretches of historical, not scientific, time. Much of the data testifies to complex civilizations so remote that they have become legendary tales and fables now unworthy of our respect or investigation.

Donald Patten points out in *The Biblical Flood and the Ice Epoch* that the Dravidians of India, a very early people, had "astronomical and astrological themes in their literature and religion. They were concerned about perishable versus imperishable worlds, about ages, catastrophes, cycles, and new ages, and they were concerned about mathematics, causes of natural phenomena, planets, orbits, and zodiacs."[280] He then reminds us that the Chaldeans, Druids, Egyptians, Germans, Greeks, Incas, and Mayas all had complicated astronomies and astrologies. Zecharia Sitchin, contemporary advocate of the ancient astronaut thesis, mentions that Alfred Jeremias, in *The Old Testament in the Light of the Ancient East,* "concluded that the zodiac was devised in the Age of Gemini (the Twins)—that is, even before Sumerian civilization began. A Sumerian tablet in the Berlin Museum (VAT. 7847) begins the list of zodiacal constellations with that of Leo— taking us back to circa 11,000 B.C., when Man had just begun to till the land."[281] This evidence contrasts sharply with the image given us by cultural evolutionists of cavemen at that time hunting on the edge of glaciers and only dimly perceiving the nature of life and death.[282]

The testimony from scholars about the knowledge of the ancients rarely reaches us in either textbooks or popular articles on the origins of mankind. Peter Tompkins, in *Mysteries of the Mexican Pyramids,* moves the dates even farther back, suggesting cultures so ancient that we can hardly imagine their existence. "Like the Maya, and their possible predecessors the Olmecs," he writes, "the Chaldeans had records of stars going back 370,000 years, while the Babylonians kept the nativity horoscopes of all children born for thousands of years, from which to calculate the effects on humans of various planets and constellations."[283] These numbers are startling and raise questions that we should have faced honestly when this data was discovered. They are studiously avoided by historians studying ancient times and not taken seriously or simply not mentioned at all.

Why would ancient peoples want to know the configurations of the heavens over such a tremendous amount of time? Did they project backward mathematically to arrive at these numbers? Or do these records represent actual observations? The question affects evolutionists and creationists alike and demonstrates that both groups have been content to let data such as this fade into the background because it is disruptive to the stories they tell. The figure of 370,000 years contradicts the creationist estimate, which favors an absurd 6,000 years, but provides no support to cultural evolutionists and archaeologists, who debate the various "lithic" ages of 150,000 to 100,000 years during which our species supposedly cringed in caves, afraid of the darkness and stars, barely capable of sharpening one side of a flat rock. If both groups deny the possibility of human existence and intelligent observation of the heavens dating back as far as Tompkins's records show, then our original accusation that science has in some ways appropriated the biblical worldview and is answering its questions holds firm.

In fact, the Western, Christian view of historical time stands out as an anomaly in comparison with other cultures, large and small. Almost universally, other people speak of a series of worlds prior to the present one, when things were entirely different on Earth, when other peoples and exotic animals were alive and prospering. In general, their memories are not fables, and contain some reasonably specific ideas that might be verified, given some openness. Depending on the tradition, people speak of "worlds" or "ages" when they are referring to the totality of the previous world, including its humanoid creatures and their social structures as well as the physical world. Other people speak of "suns" when the cosmology was different from what it is today. These memories should be included in a rendering of secular, human history of the planet.

The list of cultures that embraced the idea of past ages is astounding. A preliminary survey reveals that the Chinese have ten previous ages; the Polynesians and Icelanders have nine; and the Etrurians, the Visuddhi Magga, the Bahman Yost, the Annals of Cuauhtitlan, the Sybylline Books, the Mayas, and the ancient Hebrews all report seven previous worlds. Remembering four previous worlds are the Hesiod, the Bengals, the Tibetans, the Bhagavata Purana, the Mexicans, and many American Indian tribes. With more concentrated research, one could find considerably more cultures that believed in a multiplicity of worlds. We have no basis for rejecting their statements except to say, as many academics are prone to do, that we don't believe them.

In view of the Western/Christian propensity to support the single world theory and pretend that we are unique, it is important to note briefly that the multiplicity of Hebrew worlds since Genesis must be understood within the context of early Hebrew beliefs. Louis Ginzberg, noted Hebraic scholar and historian, in *The Legends of the Jews,* says: "Nor is this world inhabited by man the first of things earthly created by God. He made several worlds before ours and he destroyed them all."[284] In a variant Jewish tradition, according to Immanuel Velikovsky, "Several heavens were created, seven in fact. Also the seven earths were created; the most removed being the seventh Erez, followed by the sixth Adamah, the fifth Arka, the fourth Harabbah, the third Yabbashah, the Second Tebel and our own land called Heled, and like the others it is separated from the foregoing by abyss, chaos, and waters."[285] Genesis describes only the beginnings of the present world, and it is a summary of Sumerian beliefs.

Much speculation about the accomplishments of ancient peoples centers on the Sumerians. They spoke a language that was unrelated to any other that we know. We can read their language primarily because of the multilingual inscriptions that have been found and the use of Sumerian words by later peoples. Scholars have waxed eloquently about the achievements of the Sumerians and have traditionally credited these people with the birth of civilization. Some scholars even contend that the Sumerians exerted great influence on the Egyptians, who rival them in longevity. Zecharia Sitchin summarized the praise of scholars for these people: "H. Frankfort (*Tell Uqair*) called it 'astonishing.' Pierre Amiet (*Elam*) termed it 'extraordinary.' A. Parrot (*Sumer*) described it as 'a flame which blazed up so suddenly.' Leo Oppenheim (*Ancient Mesopotamia*) stressed 'the astonishing short period' within which this civilization has arisen. Joseph Campbell (*The Masks of*

God) summed it up in this way: 'With stunning abruptness ... there appears in this little Sumerian mud garden ... the whole cultural syndrome that has since constituted the germinal unit of all the high civilizations of the world."[286] These are high accolades indeed, but why?

Most scholars adopt the conclusions of the pioneering work of Samuel Noah Kramer, the reigning authority on Sumer for many years. They particularly cite his famous twenty-five "firsts" from his books *From the Tablets of Sumer* and *History Begins at Sumer*. Among the Sumerians' accomplishments are the first wheels, kilns, bricks, high-rise buildings, commercial agriculture, metallurgy, medicine, irrigation, and city planning. More important, in the social/cultural realm, they devised the first schools, courts, temples, legislatures, priests, kings, administrators, librarians and library catalog systems, historians, literary debates, poetry, music, and art. Kramer became so enthusiastic about Sumerian achievements that he later revised *History Begins at Sumer* to include thirty-seven "firsts."

Different scholars propose alternative inventors for some of these disciplines, and as we discover additional artifacts in the ruins of different societies, the emphasis shifts back and forth among contending perspectives. Agriculture, for example, could have originated in the Middle East highlands, somewhere in the Western Hemisphere, or in China. Metallurgy's origin is attributed to a variety of locations, depending on how extensively scholars compare various digs and what their dating scheme is. Finding very ancient mines in South Africa dating perhaps to 50,000 B.C. may radically change our concept of how and when people began to work ores for metals.

All these different disciplines come together in Sumer as one interwoven cultural/social complex that raises questions of origin that cannot be answered by allocating the various "firsts" to other peoples. If this complex did arrive together, then we might give serious thought to the possibility that an extraterrestrial civilization brought it to the Sumerians, that the Sumerians inherited this knowledge from the survivors of a previous world, or that they themselves were a different people than we have imagined. When we broach the subject of ancient astronauts, many scholars automatically turn their backs and return to their fictional speculations about caveman paintings. Our intent here, however, is not to *prove* the existence of ancient astronauts but to *examine* the idea.

Generally, when the idea of ancient astronauts is brought up among its adherents, they suggest seemingly impossible engineering feats and point to the strange ruins at Baalbek, proclaiming that ancient astronauts left these megaliths as evidence of their sojourn on Earth. That conclusion can hardly follow, however, from an examination of large stones. Megaliths may get us interested in prehistory, but they cannot tell us very much, and we tend to fill in the enormous gaps in our knowledge with speculations. From the multitude of people who have made claims regarding ancient astronauts, I have chosen three scholars who seem to represent possible fruitful approaches to the question: Carl Sagan, C. A. O'Brien, and Zecharia Sitchin.

Sagan assumes that the Sumerians were already living as a sedentary people with some measure of civilization. He does not elaborate on the scope of their accomplishments or the origin of their knowledge and technology. He suggests that a strange aquatic creature with a vast scientific knowledge then arrived on their shores, and they become the first modern society through his teachings. Sagan quotes fragments of legends from Alexander Polyhistor, Abydenus, and Appollodorus, suggesting that these ancient writings "present an account of a remarkable sequence of events. Sumerian civilization is depicted by the descendants of the Sumerians themselves to be of non-human origins. A succession of strange creatures appears over the course of several generations. Their only apparent purpose is to instruct mankind. Each knows of the mission and accomplishments of his predecessors. When a great inundation threatens the survival of the newly introduced knowledge among men, steps are taken to insure its preservation."[287]

With Sumer apparently founded around 4000 B.C., and Noah's flood often dated around 2800 B.C., that would mean a useful life of the Sumerian culture of only around 1,200 years. This would barely rival Egyptian longevity and would hardly account for the astronomical tables that purport to track stars for 370,000 years. Even more confusing is the fact that this creature came from the sea, which would suggest that we are dealing with some kind of amphibian, or that astronauts from other planets like to spend their quality time in our oceans. The idea of a superhero coming to people and teaching them the arts of civilization and/or simple crafts is pervasive in North America and is found in many other parts of the world, albeit without any of the fundamental changes such as occurred at Sumer. Pueblos and Navajos, for example, were much better fed after the visit of the Corn Mother, but they did not begin compiling astronomical charts and building temples. Unfortunately, Sagan did not follow up on his idea and demonstrate *how* the appearance of a being with superior knowledge changed the behavior of the Sumerian people. He was apparently content to simply suggest that an unidentified complex of institutions and technologies arrived and was adopted. Cultural evolutionists should tell us whether these changes were primarily social or technological in nature, and which came first.

Zecharia Sitchin provides such an extensive development of his idea that his thesis had to be presented in a series of books that sought to cover the period from the prehistoric cosmos to the landing of the ancient astronauts, their subsequent settlement, and their eventual disappearance. Sitchin does connect with the chronology of the Bible in his book *The Wars of Gods and Men,* in which Abraham appears as a character. Briefly, according to Sitchin, the Earth is invaded by an expedition of astronauts who come to Earth to mine gold. After working in the gold mines for hundreds of thousands of years, they rebel and demand that the elders of the expedition create a worker who can do the manual mining labor in their place. The leader of the expedition begins to experiment with genetic engineering, creates a bevy of strange animals later to appear in Middle Eastern mythology, and finally creates a hybrid from a local hominid by crossing his genetic structure with that of the astronauts.

Some hybrids work as slaves in the mines, and later some become servants in the cities established by the members of the expedition. The hybrids walk around naked since they are not believed to be fertile and apparently have no sexual urges. "Birth goddesses" who have hybrid eggs planted in their wombs give birth to new servants as they are needed. Sitchin cites some tablet scenes that feature naked humans with a clothed god or goddess and others that show goddesses lined up to give birth. If these tablets actually exist, why haven't scholars made some reference to them during the last century? The problem with Sitchin's theory is that we are never certain whether these tablets actually exist because he does not give accurate documentation and because they are so foreign to other tablets we have seen from the same civilization.

Although Sitchin doesn't state it directly, it is plain that early institutional religion is a device invented to control humans. The astronauts create the concept of "gods" to distinguish themselves from us socially. Temples are built that are actually elaborate villas for the gods. Ziggurats and the multitude of ancient, public city buildings are constructed for the administration of the cities so that control can be maintained. The prettiest woman in the city has to spend the night in a room at the top of the ziggurat in case the astronaut who rules the city stops by for some fun[288] (much like what may happen in some fancy apartment buildings in Washington, D.C., I would suppose). Things get worse and the population of humans grows dramatically.

Then the astronauts become aware that a massive physical catastrophe will occur shortly. They hold a council, find a way to protect themselves, and decide not tell humans about its approach so they will be exterminated. One astronaut does tell a human (here insert the name of your favorite flood hero), however, and instructs him in how to build a boat so some humans can be saved. The catastrophe occurs, most humans and most animals are destroyed, and we are now well into the chronology of ancient history that leads to the biblical account of the rise of the Hebrews as a distinct people. From the picture Sitchin sketches, the Noah figure seeks to save only domestic animals and doesn't scour the Earth for unicorns and so forth. He also takes a large number of people with him, according to extracanonical sources.

When we connect the ancient astronaut thesis with the biblical families, we must remember that Abraham and his people are really Chaldeans from Ur, a city that was apparently rebuilt after humans once again had a sufficient population to build cities. Being a shepherd, Abraham would probably have had just enough knowledge of the popular version of Sumerian prehistory to have passed down the stories that now compose Genesis. We are not told that he was one of the palace elite or even educated. So perhaps he could not have known of the complex histories recorded by the Sumerians regarding the flood or other past events. Considering that Genesis was written after the exodus, and perhaps even after the entrance into Canaan, by people who had heard the traditions as folklore, what we probably have in Genesis is a "*Reader's Digest* condensed version" of Sumerian history.[289] An examination of Genesis shows that it is a sequence of personal family stories. We have no assurance that they follow each other in strict descent to provide a family genealogy. Creationists who read Genesis literally to compose their chronological timelines are seriously mistaken if they assume that it contains accurate historical facts about early man.

Different aspects of Sitchin's scenario give me pause, but I have presented an expanded synopsis of his thinking because it has elements that can be used to critique some aspects of the beginnings of Judaism and

Christianity and, by extension, the Western view of history. If we have goddesses giving birth without benefit of sexual intercourse, we have a way to
understand the virgin-birth idea that was so popular in the Middle East
and see that it was a mark of honor. If the astronauts were pretending to
be "gods," there was ample reason for them to be jealous and demand obedience from their humans. If they in fact brought some kinds of flying
vehicles with them, stories about their fantastic flights may simply have
been descriptions of what humans observed. Both the Hopis and the people of India have many stories about the flying machines of the ancients,
so this thesis can become a foundation for further research.

C. A. O'Brien, English prehistorian, offers an alternative reading to the
Sitchin scenario, placing his emphasis on the garden of Eden, the events
that caused its destruction, and the eventual founding of cities in the
region known as the Fertile Crescent. Sitchin understands these cities as
earlier efforts to create a grid for landing spaceships in this area. O'Brien
suggests that many of the terms we uncritically accept as descriptive of
religious personalities were in fact offices or ranks in the ancient astronaut
social structure. Thus the YHWH was an office filled by a succession of leaders rather than a name for a specific god. He strikes a chord here because
the Old Testament has many admonitions against worshipping "other
gods," as if those deities were actual beings who could compete with the
Hebraic deity.

We are accustomed to explaining religion in terms of "worshipping" our
deities and attributing to them the most fantastic of powers. That would
be the proper stance of a creature standing before someone of infinitely
greater knowledge and technology. Carl Sagan writes, "The astronauts
would probably be portrayed as having godlike characteristics and possessing supernatural powers. Special emphasis would be placed on their
arrival from the sky, and their subsequent departure back into the sky."[290]
These themes are found in many peoples' histories. Our idea of heaven

must come from this experience. I am reminded here that the Osage Indian stories, which tell of a hybrid of "earth" and "sky" peoples, are perhaps an echo of that time.

Sitchin's and O'Brien's basic argument regarding the creation of man rests on linguistics. "The very terms by which the Sumerians and Akkadians called 'man' bespoke his status and purpose," Sitchin writes. "He was a *lulu* ('primitive'), a *lulu amelu* ('primitive worker'), an *awilun* ('laborer'). That man was created to be a servant of the gods did not strike the ancient peoples as a peculiar idea at all. In biblical times, the deity was 'lord,' 'sovereign,' 'King,' 'Ruler,' 'Master.' The term that is commonly translated as 'worship' was in fact *avod* ('work'). Ancient and biblical Man did not 'worship' his god; he worked for him."[291] This point is important because it is the firm contention of Christian apologists that the "sovereignty" of god is what distinguishes that religion from all others.

Ian Barbour says that "clearly the biblical story differs from other ancient creation stories in its assertion of the sovereignty and transcendence of God and the dignity of humanity."[292] I am dubious about how much dignity for humanity is found in Genesis, but with the ancient astronaut thesis we can certainly see where the idea of the sovereignty of god originated. It was that of the master and slave relationship. Later on the same page, after asserting the dignity of man, Barbour admits, "In the Babylonian story, humanity was created to provide slaves for the gods," forgetting perhaps that Genesis is a summary of a much more extensive Chaldean tradition and basically supports the ancient astronaut thesis. His admission that humans were created as slaves gives added credence to the ancient astronaut thesis.

When we examine Christian theology through the centuries, we find that it always uses political and judicial images. It is as if the trauma of life as slaves to superior beings could not be forgotten. (We know Protestant Christianity certainly never overcame the embarrassment of being naked.)

Barbour discusses how Christianity has phrased its explanations by using judicial models, and here again we see the trauma of being on the bottom of the social pyramid in Sumer. Barbour says that the "penal substitute model uses the images of a law court. The satisfaction for justice requires a penalty for our offences; Christ as substitute bears our punishment and we are acquitted."[293] But why would a society model its understanding of cosmic history after a human institution? Many societies modeled their ideas about the world after the family that could be seen in both animals and humans. Why would people define the purpose of history as a deity's seeking to punish humans unless they had already experienced that setting while the gods were controlling them?

Barbour offers a second model used by Christian theologians: the "sacrificial victim model," which "uses the images of the temple sacrifice. Christ as both priest and victim (as in the letter to Hebrews) provides expiation for man's sin."[294] Now, a point that Sitchin makes, although not too clearly, is the importance of these blood sacrifices of ancient times. The "temples" were apparently villas for the various gods, since they were always dedicated to a particular god. The sacrifices may well have been simply that the human priests had to cook meals for the temple god. O'Brien emphasizes that the meat had to be well cooked since the astronauts had experienced an illness from badly cooked food early in their occupation of the valley in which the garden of Eden was located. The old saying that gods loved the smell of the sacrifice could well have stemmed from a real memory that, at a good barbecue, the gods loved to smell meat cooking— somewhat akin to Lyndon Johnson savoring a Texas feed.

Other miscellaneous tidbits make sense when we consider the ancient astronaut thesis. Everyone familiar with ancient mythology must be shocked to find that in the Sumerian, Hindu, and Greek pantheons we find abundant evidence of completely dysfunctional groups. Adultery, incest, murder, betrayal, political coups, and feuds of great intensity occur as a matter of course. Grandfathers can hardly wait until their granddaughters are "of a desirable age," as they say in the American South. Without exception, the human followers who "worshipped" these gods lived a less promiscuous life than did the deities. How could any of these "religions" have flourished in human society for any period of time if these deities represented the highest expression of moral and ethical behavior that we can conceive? Who would worship Zeus and others unless compelled to do so? The gods appear to have been feasting on Viagra rather than ambrosia. Their behavior as public figures was probably not duplicated by any society until our recent Congresses.

Christian theology has colored our interpretation of ancient texts so that we tend to think in exalted, abstract terms, believing that the peoples of the Middle East thought in the same manner as we do. We can look at other materials that will shed light on the ancient astronaut interpretation. "El" seems to have been a reasonably widespread word designating the highest god in the regional pantheon. Marvin Pope, noted Old Testament scholar, looked at the Ugaritic texts and attempted to discern what the original meaning of this term might be. He found that "there is hardly anything that could be called a creation story or any clear allusion to cosmic creativity in the Ugaritic texts." But, he said, "The tradition of YHWH as a Creator God ... is a prominent feature of the Old Testament and YHWH was almost certainly identified with El. It is altogether probable that El was a Creator God, but the *Ugaritic allusions to El's creativity are in terms of generation and paternity*. El is called ~~ [citations to Ugaritic texts omitted], 'Father of Mankind' and *bny bnwt,* 'Creator of creatures.' The translation of 'Creator of creatures' for *bny bnwt* is not quite satisfactory or adequate, but a woodenly literal rendering of 'Builder of Built Ones' or 'Begetter of Begotten Ones' would be rather awkward."[295] Pope's analysis gives comfort to the Sitchin and O'Brien thesis.

What does this mean in terms of the history and worldview of these people? They apparently knew or believed that El had had some intimate involvement in the creation of human beings and that the designation "father" had a strong genetic flavor rather than a sociological/religious emphasis. Perhaps "El" was a donor of genetic material, and the "creation" of man involved him personally. This line of interpretation, which coordinates the general beliefs of the peoples of this area instead of isolating the Old Testament and pretending that the Hebrews were immune from cultural interchange with their neighbors, can lead to a radically different secular history for this part of the Earth.

Theologians, as we have mentioned, put much emphasis upon the "sovereignty" of god and tell us that all creation is subject to his powers. That belief may be a later interpretation by Hebrew thinkers. We remember from the mythology of Sumerian, Hindu, and Greek traditions that the head god, whatever his name, barely had control of the other gods and often faced quarrels and disagreements within their ranks. Pope paints the same picture from his reading of the Ugaritic texts: "The disrespectful behavior of Prince Sea's messengers, El's ready capitulation to their demands, his apparent helplessness in the altercation that breaks out in the divine assembly, his submission to 'Anat'sr threats, is rather difficult to understand, if his authority and power are really commensurate with his nominal position. It is hard to see how the Ugratitians who composed and read and heard these ironical episodes recited could have had a firm belief in El's supremacy."[296]

Most Western religious thinkers will reject out of hand the linkage of Sumerian and early Hebraic sources with their traditional interpretation of the Old Testament. It is nonsense to pretend, however, that all of the histories and references to other peoples must fade away into obscurity because we *prefer* to regard the Old Testament as historically accurate and to exclude other literature that originates in the same cultural complex. The ancient astronaut thesis opens a new area of historical research in that by synthesizing data rather than isolating it, one can construct a more realistic alternative history of mankind. We cannot, then, rely on the accounts of Genesis alone to give us accurate historical data. But taking them together with the Sumerian and other traditions, we can construct a new history of the region that has both internal and external consistency.

Let me be clear about the purpose of this chapter. With secular history, we seek to go back to modern man's earliest appearance. Yet when we do, at least in some of the ancient ruins we have uncovered, there is already a complex civilization present, with both technology and institutions clearly similar to that which we possess. We need a history that will easily and comfortably explain these great cultural complexes. The basis of Western religions is their claim of the absolute primacy and sovereignty of god as the creator. What if this belief is not an exalted abstraction that has emerged from a religious tradition but a memory of a unique event in our planet's history? Clearly, the ancient astronaut thesis spans the tremendous gulf between these two views. Western religion may simply be the historical remnants of an ancient cargo cult.

I do not believe the ancient astronaut thesis, contrary to what I'm sure many reviewers will insist. Following Occam'srazor, however, I am forced to admit that it ties up a lot of loose strings. I do believe that it has much to tell us and should be a topic for serious historical investigation rather than simply the concern of flying saucer groups.

Unfortunately, at the present time academics seem to be rewarded with advancement primarily when they keep the knowledge of the past grounded in narrowly focused specialty topics, thereby gaining an immense reputation by becoming an expert in a minuscule field defined exclusively by themselves and their colleagues. Synthesizers—those who try to paint the larger picture—do not do well in academia, and for that reason most of the truly creative work is being done outside the ivy-covered walls. Other scenarios can be brought forward that view the data in entirely different ways than I have suggested. It is in the nature of real scholarship to propose new ways of interpreting data instead of simply defending outworn theses. In that sense, and because so much is being published for the general reader, sooner or later some serious consideration will have to be given to these ideas. The other religions, then, have much to contribute to our understanding of the universe and ourselves, but they must be taken as seriously as we take the Western traditions.

CHAPTER NINE

EFFORTS AT SYNTHESIS

THE WESTERN PHILOSOPHICAL WORLDVIEW has been unraveling for some time. We might trace the beginning of its erosion to the relativity theories of Albert Einstein or the subsequent developments in quantum physics, astronomy, and biochemistry during the last half of the twentieth century. Certainly the discovery of DNA and RNA has been critically important in forcing us to look at nature anew. The plethora of wars during the past century that demonstrated that we are not yet civilized should have shaken our confidence in the old way of understanding the world. Two closely related concepts have recently come to the foreground as evolutionists and creationists attempt to resolve the question of origins. They are intelligent design and the anthropic principle. They are the contemporary world's expression of how we can think about the physical world and explain our successes in it.

It may be that the battle is starting to shift from attacks on creationism to criticism of intelligent design as more scientists see designs that cannot have been made by a gradual process of change. According to James Glanz, reporter for the *New York Times*, intelligent design has become the latest and most sophisticated version of creationism, and his April 2001 article "Intelligent Design Grows as Challenge to Evolution" seemed calculated to generate additional hysteria among the die-hard evolutionists. "In Michigan," Glanz warned darkly, "nine legislators in the House of Representatives have introduced legislation to amend state education standards to put intelligent design on an equal basis with evolution."[297] Why the idea of a rational universe would be a threat to evolution is a question Glanz did not answer. The theory does not logically lead to a biblical creator and, if properly presented, should pose no threat to the idea of random change creating a comprehensible universe. But what is intelligent design that it strikes fear in the hearts of scientists?

The argument for intelligent design suggests that if certain measurable numerical values that we have discovered in nature were but a shade different, we would not have a functioning universe at all. Victor Stenger gives the specifics of the argument: "If the universe had appeared with slight variations in the strengths of the fundamental forces or the masses of elementary particles, that universe would be pure hydrogen at one extreme, or pure helium at the other. Neither would have allowed for the eventual production of heavy elements, such as carbon, necessary for life."[298] He offers additional examples: "If gravity had not been many orders of magnitude weaker than electromagnetism, stars would not have lived long enough to produce the elements of life. Long before they could fabricate heavy chemical elements stars would have collapsed."[299] Thus stated, intelligent design reminds one of Walter Cronkite's closing remarks on finishing the news: "That's the way it is." We have a universe that we can describe mathematically and, if our figures are correct, it couldn't be otherwise. It is exceptionally well designed to function as it does.

Most articulations of intelligent design are based upon our knowledge of the larger cosmos, and we have been able to gather an impressive body of rather precise measurements, as described above. Some creationists, realizing that there is a vacuum within which additional arguments can be made that will support their goals, seek to move beyond what can logically be said. For example, Hugh Ross, an avowed fundamentalist, sought to extend the argument of intelligent design to the sun-earth-moon system. He made a list of nineteen specific physical phenomena that he believed make our solar system unique and, Ross argued, therefore "designed." The list is as follows:

1. number of companion stars
2. parent star birth date
3. parent star age
4. parent star distance from center of gravity
5. parent star mass
6. parent star color
7. surface gravity
8. distance from parent star
9. thickness of crust
10. rotation period
11. gravitational interaction with a moon
12. magnetic field
13. axial tilt
14. albedo (ratio of reflected light to total amount falling on surface
15. oxygen to nitrogen ratio in atmosphere
16. carbon dioxide and water vapor levels in atmosphere
17. ozone level in atmosphere
18. atmospheric electric discharge rate
19. seismic activity[300]

Some of these items are easily understood by many laypeople, while others are very technical. The cumulative effect of a list of technical concepts may have some weight with readers who are impressed with lists. The list is built upon a rather naïve set of assumptions and an apparent lack of knowledge of the history of science.

A number of recent studies provide data that raise serious questions about Ross's claims. Some claims are easily refutable. Albedo cannot be constant and must change somewhat in droughts, ice ages, and changes in the atmosphere. Some scholars have suggested that albedo might have been different in the remote geologic eras when conditions were radically different. Thus it cannot offer proof of design when it varies so considerably in different geologic eras. The relationship with the moon fluctuates significantly and there is a current theory, offering some evidence that must be taken seriously, that the moon was formed by a collision with a passing celestial body. At a 1984 conference in Hawaii about the moon, Donald R. Davis and William K. Hartman suggested that it was born cataclysmically from a collision between the Earth and a celestial body nearly as big as Mars: "Some four and a half billion years ago," they theorized, "the speeding wanderer collided with the early earth and shattered to smithereens, blowing debris and bits of the Earth's crust into space. The ring of orbiting debris then coalesced into the moon."[301] How, then, could the moon's present parameters be designed if it originated in an unexpected collision?

Some scholars are suggesting that both the current rotation *and* revolution rates of the Earth, cited by Ross as evidence of design, were different in former times, and the thickness of the crust certainly cannot be relevant to design theory since it varies from place to place and era to era. The magnetic field has been known to fluctuate from time to time, the major question in this respect being whether or not its fluctuations are periodic or caused by collisions with other bodies. Axial tilt has also been attributed to interaction with extraterrestrial bodies. To cite seismic activity as providing evidence of design is simply perverse. In short, we can see Ross's design only if we are not aware of continuing research now being conducted in many countries.

More important for our consideration, however, is the fact that the results of scientific experiments continue to modify our best understanding of natural processes. The histories of Western philosophy and science tell us that each generation of thinkers supports the doctrines of its day, although theories are continually rendered void by subsequent developments. Weren't the scientists and theologians of Galileo's day content that the universe, as they were able to understand and measure it, was designed for us? Wasn't the Ptolemaic system as rational as the Copernican system? When expressed mathematically, isn't the major difference the simplicity of the Copernican system? Didn't the French Academy once deride the idea that stones fell from the sky? Citing current scientific beliefs as evidence of intelligent design is dangerous in that the data is always changing, and new theories are continually being put forth to replace doctrines believed by the majority of people. Intelligent design must always be supported by our emotional response to the world, not our efforts to understand it scientifically.

Design advocates ought to restrict their arguments to areas in which they clearly have the advantage. David Foster offers some impressive figures from the field of biology: "The specificity of the haemoglobin protein is represented by the number 10 (650). What this means is that if haemoglobin evolved by chance there would be one chance in 10 (650) of it actually occurring. The specificity of the DNA of T4 Bateripphage is represented by the number 10 (78,000) so that there is only one chance in 10 (78,000) of it actually occurring."[302] Those are significant figures, and suggest that in certain areas of science, particularly at the biological micro level, the design argument does have some weight. Michael Behe argues that some processes in biology—blood clotting, for instance—are better explained by the idea of design than by chance occurrence. Design probably cannot be convincing at the macro level, and is dubious at the meso level, but can certainly be convincing at a micro level. Whether our enthusiasm for design can be maintained in view of the three different levels from which we draw evidence is another matter altogether.

Would science lead us to confusion on the question of design? Not intentionally. The physicists insist that they are describing not nature itself but our picture of nature. They assert that quantum waves are merely convenient ways of visualizing and thinking about things that we cannot see and can measure only as effects. Stenger introduces this limitation in his complaints about design. "The laws of physics," he says, "at least in their formal expressions, are no less human inventions than the laws by which we govern ourselves. They represent our imperfect attempts at economical and useful descriptions of the observations we make with our senses and instruments."[303] Does DNA research provide us with facts whereas physics give us probabilities? If so, then design arguments need to focus on biological data and leave speculation about the larger universe alone.

How can there be design when Earth history suggests that the Earth is a small planet periodically bombarded by celestial bodies of nearly equivalent size? If our Earth history is now divided into a series of catastrophic events between which there are periods of calm and uniformity, then we have not one "creation" to explain but several. How did whole biotic systems, complete with prey-predator and symbiotic relationships, arise from a planet that had been virtually destroyed by fire, flood, wind, and earthquake? Why were these succeeding biotic systems so radically different? How did some species survive and prosper when so many others did not? Intelligent design, then, may be an argument that may be applied to

organic life, but it cannot hope to explain the origin of biospheres following catastrophes.

The anthropic principle, which is the companion theory to intelligent
design, is at best confusing, considering the admission by physicists that
their formulas are descriptions of probabilities and that "natural laws" are
merely expressions of statistical observations. Brandon Carter, British cos-
mologist, coined the phrase in 1974 and, according to John Noble Wilford,
science columnist for the *New York Times*, it represents the belief that
"since life exists in at least one place in the universe, the physical laws gov-
erning the universe, the fundamental forces and particles and so forth have
to be such to account for the evolution of stars and galaxies or anything
else essential to the emergence of intelligent life."[304] This interpretation
varies from intelligent design in that it is more precise—it conceives a universe that necessarily supports the organic life we have on Earth.

Ultimately, of course, the anthropic principle analyzes the universe from
a man-centered perspective, with our species representing all of life. We are
simply gathering the data we have and using it to explain our premise—
that we are important. When the smoke clears from this twisted logic, we
find a strong biblical theme here—that the world was made for man. If we
maintain, however, that at the macro level there does not seem to be much
coherence and that a cosmic collision can radically change the forms of life
on our planet, we inevitably have to return to a modified form of deism or
theism—and in Western society, this means the Old Testament god.

The novelty of the idea that the universe was designed so it could produce life has had a certain fascination for scholars, so various versions of it
have been put forth for consideration. John Barrow and Frank Tipler, in
The Anthropic Cosmological Principle, traced the genesis of the idea back to
the Greek Anaxagoras and even sought to link it to creation stories of other
societies, including a number of American Indian traditions. They
identified the many versions of the principle that had been developed since

Carter coined the term. The most popular was "the Strong Anthropic Principle (SAP). The Universe must have those properties which will allow life to develop within it at some stage in its history."[305] SAP has three different expressions:

1. There exists one possible universe "designed" with the goal of generation and sustaining "observers."
2. Observers are necessary to bring the universe into being. (This expression is popularly nicknamed PAP, for Participatory Anthropic Principle.)
3. An ensemble of other different universes is necessary for the existence of our universe. [306]

Barrow and Tipler also identify the Final Anthropic Principle (FAP): "Intelligent information-processing must come into existence in the Universe, and, once it comes into existence, it will never die out."[307] These versions of the principle cover many contingencies and represent a good deal of the discussion involving complex hypothetical analyses of the conditions necessary to explain the presence of life on Earth or in the universe. But some conditions seem unlikely. To gauge the requirement that the universe provide for the origin of intelligent life, we should also include all the other forms of intelligence and not restrict ourselves to humans. The universe could as well be designed for birds or bears as for ourselves. Here we find a reluctance to discard the old biblical themes and approach the question of organic life in a straightforward manner.

Ian Barbour explains the anthropic principle in terms that most lay people can understand: "If the strong nuclear forces were even slightly weaker we would have only hydrogen in the universe. If the force were even slightly stronger, all the hydrogen would have been converted to helium.

In either case stable stars and compounds such as water could not have been formed. Again, the nuclear force is only barely sufficient for carbon to form; yet if it had been slightly stronger, the carbon would have all been converted to oxygen."[308] This principle has been broadly defined by its adherents and subjected to many kinds of interpretations, making it similar in content to both religion and evolution in that anyone can attach their own meaning to it.

Michael Behe, responding to some of the mathematical inquiries concerning the need for more universes, which express possibilities that are not or cannot be realized, says that the anthropic principle "states that very many (or infinitely many) universes exist with varying physical laws, and that only the ones with conditions suitable for life will in fact produce life, perhaps including conscious observers."[309] Hugh Ross says it means, "Everything about the universe tends toward man, toward making life possible and sustaining it."[310] And John Noble Wilford observed that some cosmologists in otherwise secular discussions about the universe had been stunned when their colleagues suggested that the anthropic principle be considered in determining the nature and properties of forces and matter in the physical laws of the universe, so that they could be used to conclude that humans were part of the design scheme.[311] Considering that most scientists now accept the thesis that an errant comet eliminated the dinosaurs, thereby opening an evolutionary niche for humans, it should have been expected that cosmologists would be annoyed when asked to formulate cosmological principles so as to enhance anthropic theory.

Science writer Margaret Wertheim, summarizing the argument, explained that the principle "proposes that the universe has been specially tailored, or 'fine-tuned,' to enable the emergence of life. Proponents of the principle ask one to imagine all the physical laws to which the universe might conceivably have been subject, and all the possible values that important physical constants, such as the mass of the proton, might have assumed. Yet, the proponents continue, the laws and constants that happen to hold in our universe are virtually the only ones that could give rise to a universe hospitable to intelligent life."[312]

But are the "constants" actually constant? According to James Glanz and Dennis Overbye, writers for the *New York Times*: "An international team of astrophysicists has discovered that the basic laws of nature as understood today may be changing slightly as the universe ages."[313] We might argue that aging is also designed, or we might admit that, again, whatever expressions we use are only projections of what we want to find. Margaret Wertheim offers a sensible critique of the principle: "Its proponents assume without question that the mathematical laws and constants applicable to the world around us represent a small subset of the laws and constants that could conceivably exist."[314] Here I think we deal with the problem of misplaced concreteness again. The principles we have discovered are not necessarily subsets. They could be simply the easiest way to express the relationships that we can presently observe. With the addition of more mathematics and geometry, with even better measuring devices, we can find increasingly sophisticated ways to express ourselves. Natural law, like common sense, must be revised again and again to follow in the wake of our discoveries. Additionally, if fields, as we understand them at present, are the major actors or entities in the universe, there could be innumerable ways of understanding how they relate to each other, each one of which might constitute a unique set of natural laws.

What do we do about the solid numbers, the speed of light and the mass of the subatomic particles, and so forth? Already we are seeing controlled experiments where, under certain conditions, the speed of light can theoretically be surpassed, raising questions about the entire intellectual edifice we have erected in the past century.[315] These values will probably always hold whenever we try to understand the universe in terms of weighing and measuring, and they are "good" or "true" in that context. But we see no such qualification in the Arkansas court's definition of science, nor could we get most scientists to admit this restrictive condition. With experiments returning strange data and measurements that threaten to undermine our belief in certain measurable constants, how can we possibly advocate the proposition that the universe must have been designed to produce our kind of intelligence?

Neither intelligent design nor the anthropic principle can provide a basis for replacing the current belief of secular science that chance alone has produced what we can see and measure. At the micro level we can find the strongest evidence for design, but we can respond to that contention by the Cronkite reply—"that's the way it is." But there are other current efforts to provide a new synthesis, though not as closely supported by mathematical speculations, that should be recognized and discussed. John Hadd, inclining toward the biblical creationists, urged that "a biblical religious orientation be reflected in the overhauled world/life paradigm," while conceding that "the demonstrable presence of an Intelligent Creator does not immediately equate with the God of the Bible. Nor does the presence of an Intelligent Creator automatically illuminate Purpose, either of the Creation or of human life."[316] Hadd claims there is a place for other traditions—"all cultural regions of the globe can contribute to a properly revamped world/life paradigm"[317]—but fails to define how such a paradigm can be brought about. Which traditions should surrender what concepts and practices? How can we tell whether the different cultures were "right" or "wrong," and what do these words mean? Could not each culture be correct in general terms but wrong on evidence? Or even the reverse: having good facts but reaching obviously wrong conclusions? Hadd fails to offer an example of how such a new paradigm could be constructed.

Harold Booher, in *Origins, Icons, and Illusions,* offers his view of how a new model explaining the universe could be constructed. It is avowedly a creationist tract, but worth considering. His model has five major components:

1. a first cause, which over a period of time brought into being the universe and all living things
2. an introduction, sometime after the original creation, of the second law of thermodynamics, which affects the universe, including life, on a path from low entropy to high entropy
3. a special creation and division of life into separate "kinds"
4. an extra-special creation of humankind
5. a general ordering of the fossil record of "kinds" in the geological record from simple to complex, due to catastrophic events[318]

It is difficult to see how such a model would not inflame scientific minds and produce an immediate rejection. The points are not synthesized but rather seem to be ad hoc concerns based more in emotion than logic. Once we have accepted some special interventions to create organic life and human beings, we have forfeited any philosophical standing to pursue the inquiry. Human beings would be outside the natural processes, and there would be no good reason to argue that we have any relationship with nature at all. Even in Genesis, man and other forms of life are created together. It is the garden that separates man from the animals.

Booher obviously wants a corrective revision of the manner in which we view the other organic inhabitants of the world. "Kinds" is a much better concept than species since, as we have already seen, the identification of species relies primarily on what the scientist, or graduate student, sees, and not what is present in nature. Both creationists and evolutionists fail to consider the biotic context in which organisms occur, focusing instead on individual organisms and projecting from them to the whole biotic system. We know from paleontological studies that there are sets of organisms that occur in the fossils together. Our major problem is in bringing together those organisms that might have been contemporaries but that are classified in different geologic periods and eras.

David Foster offers a scenario in which mind, under the name of logos, is the primary actor in the creative process. That suggestion should be welcomed by many theologians but would be seen by scientists as a massive invasion by theology. Its resemblance to Greek thought and Johannine theology is too obvious. Mind would, however, be compatible with many streams of Eastern thought. If we simply attributed intelligence to a self-operating universe, which is what string theory implies, then we would at least have a focal point from which we could depart. Mind/logos is also compatible with the views of those physicists who conclude that mind is to be favored over matter in the physical equations. So it has something to recommend itself as far as offering a way to reorganize our orientation to the problem of origins. Foster offers the following scenario for a creation model:

1. Logos manifests creative thought, which curves space to create matter such as those 1,080 hydrogen atoms, which then partly condense into stars.

2. The created matter has around it the corresponding curved-space field produced by its mass (general relativity theory).

3. The curved space then maintains the focal mass, which, as it were, is a sort of logos-memory.

4. Logos withdraws from the situation, having acted only as an initial creative catalyst.[319]

I fail to understand why this intelligence would withdraw from everything, leaving a material universe without ensuring that the various forms of expression, from galaxies to microbes, would also have intelligence or some semblance of it. The withdrawal would imply a Newtonian clockwork universe that could operate on its own after its creation. Strangely, this scenario seems to fit a curious phenomenon found in American Indian traditions. Many scholars have remarked that after creation, for many tribal traditions, the creator more or less abandons the world he or she has made, and the Indians thereafter must deal with spirits. This scenario would definitely not be popular with Americans who rely on god for a balanced emotional life.

The logos scenario also conflicts directly with our present observations about the universe. There are some facts that are often not frankly admitted by anyone. What do our observations tell us? "Red giants, white dwarfs, pulsars, novae, supernovas, and black holes all have one thing in common," says Harold Booher. "They are the dying remains of a once greater cosmos. The only evolution they represent is a degenerative one—one of extinction. The interpretation of exploding and collapsing stars is most useful in showing the enormous complexity that existed in the past, not in the processes leading from creation of space, time and matter to this observable complexity."[320] In view of these observations of the physical universe, how can we continue to advocate an evolutionary universe? On the other hand, what useful purpose would logos serve in creating a universe that was bound to degenerate, since the concept would not be useful *except* in answering the question of origins?

Foster seeks to demonstrate the validity of the concept of logos by an appeal to biology and modern communications theory. "In modern biology," he writes, "we have seen that beyond biochemistry we come to a world of information and literary logic in the DNA, and if one wishes to enquire 'what is behind the DNA?', there is little choice but to propose a similar logos. It would seem that the developments in biology are even more suggestive than those from physics in confirming that 'the stuff of the world is mind-stuff.'"[321] William Demski, currently one of the most prominent scholars supporting intelligent design, reduces the question to one of information, supporting Foster's idea. "For there to be information, there must be a multiplicity of distinct possibilities any one of which might happen," explains Demski. "When one of these possibilities does happen and the others are ruled out, information becomes actualized."[322] The universe, at any moment then, would be a complete thought. But who would think it? Here the various religious traditions would insist that their version of the deity was the active source, hampering our chances of reconciliation and synthesis.

Most Western thinkers forget that we begin our scientific search to understand the universe with the mathematical measurements of physical matter. In seeking to understand what we are doing at both micro and macro levels, we begin to substitute observations of matter in place of measurements. We eventually engage in pure speculation, supported by complex mathematical geometries, until we arrive at the proposition that matter does not exist except as fields of possibility. This search is comprehensive, if only from the point of view of a material universe, since we can measure the temperatures of stars. But Rupert Sheldrake's complaint still rings true: "Sounds, smells, colors, and feelings are nowhere to be found in mathematical physics because they are excluded from the start."[323] Do we then suggest a universe organized and held together by aesthetic expressions? Here we would return to the ancient Greeks and the harmony of the spheres. It may be that physics is already on this path, for we hear physicists talk about the elegance of mathematics and see them search for the god particle because it would enable them to express a logical totality that is presently waiting to be completed.

David Bohm has come the closest to resolving the problem that we confront. His implicate order suggests that mind and matter are two different projections of a single deeper reality.[324] In the sensory world they must coexist in some fashion. The problem with this conclusion is simply that we seem to apply it only in our understanding of the physical universe. Western culture, in particular, denies that anything we experience can exist without a physical cause. If mind produces intelligent design and indeed may not need a physical manifestation, then within the universe everything has a reason for being and a conscious life of its own. We may consist of a physical body and a nonphysical aspect that has been called a "soul" in popular discourse. This possibility, of course, opens the door to concepts such as reincarnation and communication with spirits, which are irrationally regarded as anathema by Western peoples.

At any rate, in intelligent design we are led not to the god of the Bible but to the variety of understandings of our experiences represented by the non-Western peoples. As a rule their rituals, ceremonies, and practices represent nonmathematical procedures to establish communications and relationships with the larger universe. We may begin our scientific search for understanding by objectifying everything, but when we reach our conclusions we understand that in the last analysis, we have subjectified things instead. We must eventually accept the proposition that everything is alive—insofar as mind is inherent in them—and that means that our science has not explained anything except the materialistic form in which life exists. We are not much closer to any kind of reality than when we began. Most important, we are confronting not empirical realities but the products of our own minds.

Fortunately, the "universe-as-object" attitude is breaking down within Western thinking. Scientific studies are showing that animals, which we formerly considered creatures of instinct alone, have mental capabilities and knowledge that we did not think possible, expanding our conception of the different forms of life. John Yaukey, in an article in *The Denver Post* titled "Animal Intellect Gaining Respect," reported, "While the conventional wisdom on animal intelligence has swung back and forth over the past

150 years, scientists believe it's now on a permanent march toward a greater appreciation of how animals naturally think rather than how much like humans they can be taught to think."[325] Yaukey explained further: "Studies of everything from chimps and elephants to whales and porpoises indicate they are capable of untrained thought well beyond mere instinct and can often communicate in high levels of detail. It's not so much that animals are performing more elaborate tricks; rather, they're showing signs of reasoning rather than conditioning, such as playing tricks and hiding tools."[326]

These studies are regarded as major accomplishments considering that our standard approach has been to train animals rather than relate to them, and to assume without further inquiry that they are incapable of thinking or enjoying complex emotions. An example of this new approach to animals can be seen in studies done by Irene Pepperberg with her parrot, Alex. Pepperberg, listing Alex's accomplishments, said that "he could identify fifty different objects and recognize quantities up to six; that he could distinguish seven colors, five shapes, and understand 'bigger,' 'smaller,' 'same,' and 'different'; and that he is learning the concepts of 'over' and 'under.'"[327] While this study contributes to our understanding, a deeper question might be asked about whether the parrot experiences the world in categories other than bigger and smaller and under and over. Yaukey says that recognition of the intellect of animals is "driving some researchers out of the lab and into the wild for clues about how animals develop the thinking skills essential for survival rather than the mental acuity to win a treat."[328] We can applaud the triumph of empirical observation over doctrine in this case.

Some critical questions need to be posed to people using this approach to verifying animal thought processes. Are these abilities of the parrot not phrased in human terms of measurement and relationships? In other words, don't we have such an intrusion of the scientific method here that the results are predictable? And over, bigger, smaller, and so forth are not concepts that we often think about ourselves except in earliest childhood when we are learning vocabulary. Certainly, animals in a natural state will gauge these kinds of spatial relationships intuitively as they holistically move through the physical environment. I suspect that these experiments, in which animals must relate to a scientific form of communication, whether or not there is a reward available upon successful completion of tasks, treat animals with disrespect and preclude them from communicating with us on their terms.

Observing the world in its natural state, refusing to alter the conditions of the meso-level world, provides people with accurate knowledge of things in themselves. It also reveals the existence of an extensive area of mind manifesting itself in animals that most Western peoples have not suspected. Accounts of instances in which there is a clear indication of the operation of sophisticated mental processes in animals are usually met with scorn and accusations of superstition and naïveté. Most Western people, particularly scientists, refuse to consider this possibility on wholly doctrinal grounds.

Charles Eastman, a Sioux Indian, related a tale of the behavior of a coyote that would certainly qualify as evidence of animal intelligence. While out hunting, his uncle had killed two deer, dressed them, and hung the meat in a tree; after a tasty meal, he laid down by his fire to sleep. "I had scarcely settled myself when I heard what seemed to be ten or twelve coyotes set up such a howling that I was quite sure of a visit from them. ... I watched until a coyote appeared upon a flat rock fifty yards away. He sniffed the air in every direction: then, sitting partly upon his haunches, swung around in a circle with his hind legs swung in the air and howled and barked in many different keys. It was a great feat! I could not help wondering whether I should be able to imitate him. What had seemed to be the voices of many coyotes was in reality only one animal."[329] It was an animal with an admirable power of abstract thought and the ability to formulate a complex plan of attack that would have done a scientist proud. Eastman's uncle admitted he could not have performed the same feat.

If we now begin to look at the universe as mind manifesting itself in material form, our understanding of our relationship to the natural world changes radically. We do not have inanimate, lifeless things anywhere. How far are we willing to go in applying this principle? Ian Barbour, explaining the physical world, said: "Inanimate objects such as stones have no higher level of integration, and the indeterminacy of the atoms simply averages out statistically. A stone has no unified activity beyond the physical cohesion of the parts."[330] Here he would be criticized immediately if he were in some non-Western cultures. They would argue that because he is using a materialistic science that cannot describe activities of the mind, he cannot understand stones. But other people can.

Many American Indian tribes, and I would suspect other tribal peoples around the globe, have ongoing intimate relationships with stones. They use them for healings, to perform errands, and to make prophecies, among other things. Sacred stones are unique in shape, easily recognized by medicine men, and found in particular places. Spiritual leaders claim that the stones talk to them. Chased-by-Bears, a Sioux medicine man, explained what sacred stones were and why they were important: "The outline of the stone is round, having no end and no beginning; like the power of the stone it is endless. The stone is perfect in its kind and is the work of nature, no artificial means being used in shaping it. Outwardly it is not beautiful, but its structure is solid, like a solid house in which one may safely dwell. It is not composed of many substances, but is of one substance, which is genuine and not an imitation of anything else."[331]

The power and capability of the sacred stones are illustrated many times by Frances Densmore, writing about Teton Sioux music. The most impressive, in my mind, is the story of Goose, a medicine man and a trader. A white trader had been criticizing the medicine men for a long time, calling them charlatans and sleight-of-hand magicians. Densmore says: "Goose entered into conversation with the trader on the subject, who offered him 10 articles, including cloth and blankets, if he would call a buffalo to the spot where they were standing. Goose sent both the sacred stones to summon a buffalo. The trader brought his field glasses and looked across the prairie, saying in derision, 'Where is the buffalo you were to summon?' Suddenly the trader saw a moving object, far away. It came nearer until they could see it without the aid of the glasses. It was a buffalo, and it came so near that they shot it from the spot where they stood."[332]

What do we make of this data? There had not been a buffalo in the western South Dakota region since 1883, when the last herd was hunted to extinction. Densmore most probably was not a firsthand observer of the event, but it was still fresh in the memories of people who had witnessed the incident. Do we have superstition here, or a new understanding of the stones based on the demonstration that they were capable of things we had not believed before? When Westerners scoff at stories such as this, what they are saying is that they personally have never witnessed anything similar. To the brush-off announcement "Stones don't do those things," the medicine men can reply: "They don't do them for you!" In theory, the implicate order provides a context in which these stones can be understood.

CHAPTER TEN

THE ROCKY ROAD AHEAD

THROUGHOUT WESTERN HISTORY, from the rise of Christianity until the present, we have believed that our solar system is a calm, well-ordered, self-functioning unit, with a sun and a decent family of planets revolving about it. We have had few reasons to suspect that the functioning of our present solar system is simply an interlude between devastating events of cosmic dimensions. Apart from meteorites (which were once bitterly rejected by the scientists of the day) and occasional comets that lit up the skies and entertained us with their brilliance (as well as offering dire omens of social disruptions with their appearance), there was a general belief that what we saw happening in the heavens had always been that way.

In 1950 Immanuel Velikovsky published *Worlds in Collision*, in which he reviewed the folklore of many different societies that dealt with the belief in the periodic or sporadic destruction of the Earth. Velikovsky introduced the heretical notion that during historic times that humans could observe and remember, comets had possibly run amuck in the solar system, coming close enough to the Earth's orbit to disrupt its period of revolution and wreaking havoc on geological strata. With the approach of these celestial bodies, mountains were pulled from the ground or thrust deep into the Earth by inconceivably massive forces. In various places on the globe, millions if not billions of organisms were buried. Traditionally, under uniformitarian constraints, we attributed all change to the gradual incremental forces of deposition and erosion, even though massive differences in the thickness of strata indicated a radically different process of sculpting the surface of the Earth.

Today Velikovsky must be sitting on a cloud somewhere chortling pleasantly as he reviews the enormous changes in scientific thinking that have occurred over the past half century. One can hardly read the science section of any of the major newspapers without finding an article on large asteroid/meteor/comet impacts on the Earth in ancient times. Some stories hint of massive collisions, or near-collisions, between planets to form the moon or to disrupt the orbital spin of the major planets, enabling them to add or discard satellites, and possibly to have created the solar system as we observe it today. We have crossed a threshold, and thinking about cosmic catastrophes is no longer forbidden. We have arrived at the beginning stage of a paradigm shift, where we are content to read stories of possible impacts that began or ended the large geologic eras without understanding the radical nature of change that these studies, if verified, demand in our view of the larger cosmos and Earth history. So chaotic is the present situation in this respect that we have not taken any major steps to contemplate the massive reorganization of our data and doctrines that now faces us.

What differences do the new data make in our scientific and religious folklore and beliefs, our dogmas and doctrines, and our conception of ourselves? How old is our planet? Do species evolve and then decline? Or are they suddenly present in the geological strata and then, within a few weeks after a comet impact, reduced to a pile of bones buried under the debris of the tsunamis caused by the collision? What caused geological mountain building or prolonged volcanic eruptions, if not the near approach or collision of another celestial body with the Earth? If there are several creations of biotic life, if other civilizations have lived before us, what credence do we give to the story of Genesis? If cosmic collisions have occurred, when is the next one due? If we have been paying attention to what is being discovered in the sciences today, when are we going to see the changes in our educational systems and philosophical worldviews that must certainly be necessary to explain who we are and what part we might play in the cosmic drama?

Today, when new scenarios are put forward and tested by a variety of thinkers, we must demand that our intellectuals offer us probable storylines that make sense of the new data. We cannot afford to live complacently with the belief that we are a favored species and will be saved from some future catastrophe by a merciful god, by the "space brothers" from flying saucers, or even by modern science. In concluding this book we can identify areas of concern where almost certainly science will have to provide us with new understandings of the world. The changes we need are so substantial that it may take many decades to fully absorb the implications of the data we have recently acquired. If any of these ideas or perspectives had been fully discussed in the Arkansas and Louisiana courtrooms, we would not now have the entrenchment of this secular evolutionary dogma that demands our allegiance. Instead we would have freedom of thought and expression pervasive in our educational institutions, and we would be encouraging the next generation of students to find the answers we need to make sense of our lives. Let us then examine some areas where new views must inevitably develop, and offer some suggestions for the future.

The erratic solar system. Far from the routine clockwork described in the work of Sir Isaac Newton, in which the solar system functioned without change of any kind during the life of our planet, we now have hints that our solar system has experienced profound catastrophic disruptions throughout its lifetime. It may indeed have been born in some major chaotic event. John Noble Wilford, reporting on studies done of Pluto and its satellite Charon, indicated that "water ice covered much of the satellite Charon's surface and is absent on Pluto; in fact, judging by their distinctive light signatures, the two bodies have completely different surface compositions. The astronomers said this supported the hypothesis that the Pluto-Charon system formed out of the shattered remains of a collision sometime early in the solar system's history."[333] A collision in the solar system? Twenty years ago this kind of speculation would have been impossible.

William K. Hartman and Donald R. Davis have revived an old hypothesis that the moon is composed of materials rejected from the Earth in an early collision.[334] "Some four billion years ago," they theorize, "the speeding wanderer collided with the Early Earth and shattered to smithereens, blowing debris and bits of the Earth's crust into space."[335] The suspected planet was estimated to be three times as massive as Mars. "The research indicates [that] an 'oblique impact' between Earth's crust and the ancient planet vaporized the upper portions of Earth's crust and mantle, spraying the material into orbit to form a gaseous disk. The ring eventually condensed into a string of small, hot moonlets that eventually coalesced into the single large moon we see today."[336] Zecharia Sitchin, incidentally, supports a similar theory; however, he contends that it explains the birth of the Earth as we know it today. Surely this theory should be attractive to proponents of continental drift since it provides an energetic mechanism by which we can explain the movement of continental plates.

Hartman and Davis estimate the date of the collision as 4.5 billion years ago. How they arrive at this date is a mystery. More than likely the date reflects a tendency of scientists to date collisions in the remote past, where everything is speculation anyway, thus avoiding the critiques of their peers.

In their book *When the Earth Nearly Died*, D. S. Allan and J. B. Delair mention an intriguing fact that relates to this idea: "A baked clay cylinder-seal made by the Akkadian civilization of Mesopotamia and preserved as specimen VA/243 at the Vorderasistsche Abtelling of the State Museum in eastern Berlin depicts the solar system as known to the Sumerians in the third millennium B.C. It shows eleven globes encircling a large rayed star representing the Sun."[337] Could one of these globes have been the "rogue planet," as Hartman and Davis call the visitor, in which case we could tie things together neatly? Or do we look for a catastrophe, other than the speeding wanderer, that eliminated one of those globes? Has our solar system been constructed out of several planetary collisions, or did it coalesce from gaseous rings around the sun, as we have been taught?

Asteroid/meteor/comet impacts on Earth. The popularity of impact craters cannot be denied. Robert S. Boyd, in a 1998 newspaper article, surveyed the field. "About 160 craters—some hundreds of miles wide—produced by collisions between Earth and high-velocity objects have been spotted so far, and researchers are turning up three to five more each year."[338] We have, in fact, so many craters identified now that scientists can begin to search for larger patterns and estimate rather closely when an object broke up, as we saw happen with the comet Shoemaker-Levy 9, and produced multiple craters on our planet. In 1996 a team of scientists discovered a chain of impact craters in the African country of Chad "that suggests ancient Earth may have been hit by a large, fragmented comet or asteroid similar to the Shoemaker-Levy 9 comet that slammed into Jupiter in 1994."[339] Scientists in Canada examining the Rouchechouart crater in west-central France found rocks they dated at 214 million years old. The date was identical to the age that had been assigned to rocks found in Quebec's Manicouagan crater.[340] Robert S. Boyd also reported that an international team of researchers had found a line of five craters "stretching 2,766 miles from the Ukraine through France into Canada and North

Dakota. Ranging from 6 to 60 miles across, the craters appeared to have formed about 214 million years ago, perhaps within the span of a few hours."[341] Do we need the 214 million years to accomplish this feat or could this whole episode be dated much closer to our time?

The comet Shoemaker-Levy 9 must have inspired scientists to look for chains of craters instead of isolating one crater and pretending it was unique. Another probable incidence of multiple craters, found in 1998, was "a train of eight craters that follows a nearly straight line across the center of the United States from Illinois through Missouri to Kansas. NASA geologist Michael Rampino believes they might be the wounds left by a comet or asteroid 320 million years ago."[342] Considering that the dates are at best estimates, we might be able to link them with a crater called the Alamo, found in Nevada. A U.S. Geological Survey announcement dated the crater to 370 million years ago; it was supposed to have occurred three million years before the catastrophe that ended the Devonian era.[343] Splitting hairs on estimated dates in the neighborhood of 370 million years ago is virtually saying that the hits were simultaneous. How can scientists distinguish this closely when both numbers are mere speculation? Would not this string of craters have closed the Devonian in many places on Earth?

Some individual impact craters appear to be solitary events and deserve attention. Geologists from IKU Petroleum Research in Norway discovered a large meteor impact site during the summer of 1999. "The 'hole' is 40 km in diameter and is evidence that a giant from the asteroid belt entered the earth's atmosphere and struck right off Norway's northern coast."[344] In 1998 a crater was discovered in Iowa near the town of Manson. It was dated at seventy-four million years old, and is the second-largest crater in the United States and the fifteenth largest in the world, according to Ray Anderson of the Iowa Department of Natural Resources. "All the dinosaurs in the central part of the United States would have been killed by that blast," Anderson reported. The meteorite was traveling at 60,000 miles per hour when it hit. It burrowed about three and a half miles into the Earth, and as it did, the sides of the crater were lifted one and a half miles above the Earth's surface.[345] American dinosaurs, then, were already extinct when the Chicxulub body hit the Yucatan.

The fourth-largest crater in the world was found in Australia in 2000 and was seventy-five miles wide, making it much larger than the Manson crater in Iowa. Amazingly, the Australians did not offer a fictional date for this event but simply noted that the age was "up for grabs."[346] This impact site does not figure into scientific calculations attempting to identify the sites where geologic era–ending and organic life extinction took place. Until such sites are linked to the geological boundaries that begin and end eras, we must consider them as free agents that changed the geology of regions but did not necessarily affect the biosphere of the whole planet at the time.

Previous disruptions. A puzzling aspect of the fascination about celestial impacts is the fact that astronomers had already seen an errant comet interact with one of the planets—Jupiter. Allan and Delair and Ignatius Donnelly had all pointed out the activities of Lexell's comet in the 1770s, which encountered the Jovian system. Donnelly reported: "In the years 1767 and 1779 Lexell's comet passed through the midst of Jupiter's satellites, and became entangled temporarily among them. But not one of the satellites altered its movements to the extent of a hair's breath, or of a tenth of an instant."[347] Allan and Delair noted: "Lexell's comet of 1770 and Brook's comet of 1889 both actually passed through the Jovian satellite system and almost grazed the surface of Jupiter and split in two."[348] Why then was the Shoemaker-Levy 9 encounter such a novelty? Why did it attract the attention of every astronomer when several comet-planet encounters were already known to have happened?

The best explanation is that suggested by Thomas Kuhn in *The Structure of Scientific Revolutions*: "Can it conceivably be an accident, for example, that Western astronomers first saw change in the previously

immutable heavens during the half-century after Copernicus's new paradigm was first proposed? The Chinese, whose cosmological beliefs did not
preclude celestial change, had recorded the appearance of many new stars
in the heavens at a much earlier date."[349] In other words, we often see what
we expect to see and are hardly aware of the world as it really is. This tendency is better seen in romantic encounters, but it does give us pause when
we realize that scientists insist that they are capable of objective, unbiased
observations and we know they aren't. It should be noted that any future
effort at understanding the heavens must be filtered through this
acknowledged handicap.

The era-ending catastrophes. Each of these impacts had devastating
effects on the geology of a region and certainly exterminated any forms of
life in the neighborhood, burning or burying millions of organisms and
creating the fossil beds we see in the geological strata today. But we have
not even discussed the era-ending impacts that seem to mark the change
from one geologic era or period to another. If you read the textbooks
explaining evolution, you will find no good explanations of how any of the
distinctive biospheres of the respective eras came to be. Instead we are told
that a disaster ended the Permian and then a new era opened. The new era,
strangely, had a biosphere in which all the organisms were clearly more
complex and more easily distinguished from what we had seen before.
A review of the nature of era-ending events is interesting.

Wandering into this thicket is hazardous because everything is in flux.
Dates suggested by one scholar are soon amended by other scholars, after
succeeding or simultaneous dating based on different data requires moving dates forward or backward. Relying primarily on newspaper accounts
of studies released by scientists, we will simply refer to current popular
explanations, realizing that they are but temporary stopping places on the
road to understanding the history of the Earth. "There are almost no fossils, and thus no evidence of biological extinction, in Precambrian strata,

but the detailed fossil record of 570 million years since the end of the Precambrian gives evidence of five great mass extinctions and about five smaller ones," writes Walter Alvarez in his book *T. rex and the Crater of Doom*.[350] These events should be regarded as major impacts because, as far as can be determined, significant percentages of biotic life are extinguished in them.

"The mother of all extinctions [which would be the Permian], which wiped out 90 percent of living species, happened about 245 million years ago. Paleontologists say other mass extinctions occurred about 214 million, 360 million, and 440 million years ago," according to Robert S. Boyd.[351] The dates suggested by Boyd can be organized as follows: disasters ending the Silurian at 440 million years ago, the Devonian at 360, the Permian at 245, the Triassic at 214, and the Mesozoic at 66 million years ago. The famous K-T boundary occurs between the Cretaceous and the Tertiary. There are supposed to be minor catastrophes between the Middle and Upper Jurassic and at the end of the Eocene epoch.[352] Like the classification of species, these figures are estimates and are dependent on the personal preferences of the individual scholars. Yet they do provide us with a historical framework for further inquiry that is more comprehensive than the restrictive homogenous scenario we are presently using.

Extinction of species. Of major importance in these disasters that appear to separate the various geologic periods is the appalling loss of organic creatures in each of these events. The Permian events eliminated 95 percent of ocean-dwelling species and at least 70 percent of land-dwelling vertebrates.[353] The Mesozoic catastrophe that ended the dinosaur occupation extinguished well over half of all living creatures.[354] Alan Hildebrand, Canadian astronomer, writing on the K-T boundary extinctions, suggested that impacts have strongly influenced terrestrial evolution, hardly a radical observation. He described the effects of impacts as "culling terrestrial species episodically. Between impacts, the Darwinian principles of survival of the fittest govern evolutionary success. We do not yet understand if evolution is hastened or retarded by episodes of catastrophic mass extinction … but the course of biological evolution on our planet has definitely been changed."[355]

"Culling" used to be a process of winnowing out inferior animals from an otherwise superior stock. Losing organisms in these high percentages is something more than culling. When we are losing the vast majority of species in one of these events, that is more than a simple reduction of inferior creatures; only the fortunate few who were in the right place survived, and they were not necessarily the right species to create a full branch of organic family trees. Survival of the fittest, in a catastrophic context, is actually survival of the luckiest. Organisms on one side of the world could thank god (or Darwin as the case may be) that the comet hit the other side of the planet and that they were far enough inland to escape the tsunamis resulting from the impact. How a shell-shocked organism could evolve quickly enough to deal with the immediate devastated landscape or a new configuration of an ocean bed is a problem that we have not even considered. With a substantial percentage of species lost, constructing Darwinian "trees of life" seems futile.

Changes in geologic time. We are accustomed to hearing that millions of years are required for geological changes. The old geological timescale currently prevails even when scholars are discussing the effects of a comet collision that would have stacked all kinds of new strata on top of each other and brought about enormous changes in wind patterns and ocean currents, creating new rivers and lakes. In July 2000 a report in *Geology* suggested that the end of the Permian era might have lasted less than 8,000 years.[356] A May 2001 article quoted Peter D. Ward of the University of Washington, who had analyzed rock samples for evidence of the shift in carbon levels at the Triassic-Jurassic break, as saying that "the extinction actually happened in about 10,000 years, a brief moment in geologic time."[357] On May 17, 2002, *Science* published a study by Paul E. Olsen, a professor of Earth and environmental science at Columbia University's Lamont-Doherty Earth Observatory in Palisades, New York, suggesting that a meteor strike could have created the conditions necessary to allow the dinosaurs to make a quick rise in prominence. According to estimates, it took only 30,000 years for the dinosaurs to become the dominant group on the planet instead of the millions of years previously thought.[358] Considering the magnitude of some of these events, the shorter time span sounds more realistic. If these dates hold and are supported by similar studies examining the other geologic eras and periods, we may need to reduce the geological timescale by a significant percentage, in effect bringing these cosmic impact events much closer to the present and perhaps even within a timescale that would include humans or their predecessors.

The majority of scientists will balk at such a reduction in the probable time span of geologic eras, arguing that radiometric measurements coincide with the old timescale. That possibility should come as no surprise since present efforts are devoted to coordinating the results of testing with longstanding beliefs. All forms of radiometric dating, however, depend on the assumption that we know the initial conditions of a rock or organism and that we are assured that nothing out of the ordinary has occurred between the initial state and when we test the materials. With comets heating rocks to super temperatures, carbon levels fluctuating wildly, and other phenomena rapidly changing the world, giving absolute dates is absurd.

Humans and catastrophes. Velikovsky was called a madman for suggesting that humans had observed radical celestial events and that folklore regarding catastrophes and "worlds" destroyed by disasters actually stemmed from distant memories of our ancestors. Now things are changing dramatically toward that point of view. Robert S. Boyd reported that researchers are now examining the possibility of lesser catastrophes having occurred in the not-too-distant past: "At least five times during the last

6,000 years, major environmental calamities have undermined civilizations around the world. Some researchers say these disasters appear to be linked to collisions with comets or fragments of comets like the one that broke apart and smashed spectacularly into Jupiter five years ago this summer."[359] He suggested dates of 3200 B.C., 2300 B.C., 1628 B.C., 1159 B.C., and 535 A.D. The precision of these dates can be questioned, but they are roughly consonant with a Velikovskian scenario and such familiar biblical events as Noah's flood, the tower of Babel, the exodus, and Joshua's prolonged day when the sun stood still. These stories of calamities may eventually be considered historical accounts rather than elaborated folklore.

If we are on the verge of validating the biblical stories, what about the immense body of material contained in the traditions of other cultures? How far can we go in examining some of their supposedly outlandish scenarios? An extreme example might illustrate the possibilities. Recently there has been a great debate about a time in Earth history when the planet was completely frozen. Newspaper headlines, sensationalized as usual, refer to this debate as one involving a "popsicle planet" or a "slushball planet." For the moment, it appears to be a debate between scientists at Harvard and scientists at UCLA. Paul Hoffman of Harvard, examining multiple glacial deposits in the Neoproterozoic, suggested that the Earth "lurched from icehouse to greenhouse and back again four or more times between 760 million and 550 million years ago."[360] His colleague Alan J. Kaufman attributed this fluctuation to a lack of carbon dioxide that caused surface temperatures to drop precipitously. The Earth was then rescued by "a huge volcanic eruption, a comet impact, or the sudden release of frozen methane deposits in the ocean floor."[361] Representing a less spectacular scenario are Bruce Runnegar of UCLA and David Evans of Caltech, who prefer the "slushball" or "snowball" condition that would admit land glaciers but stop them short of the tropics. Evans believes the icehouse to hothouse swing has happened only twice.[362]

The frozen planet theory might have considerable support from the Hopi tribe of Arizona. Their first world was one of infinite space, and some people say it was barely a material world. Their second world, Tokpa, ended in this manner: "The twins Poqanghoya and Palongawhoya [who were sent by Spider Women to the North and South Poles, respectively, to supervise the Earth's rotation] had hardly abandoned their stations when the world, with no one to control it, teetered off balance, spun around crazily, then rolled over twice. Mountains plunged into seas with a great splash, seas and lakes sloshed over the land; and as the world spun through cold and lifeless space it froze into solid ice."[363] What is so interesting here is that Gregory S. Jenkins, a professor of meteorology at Penn State, accepts the frozen Earth but "does not have a detailed explanation of how the Earth would have righted itself to its current 23.5 degree inclination."[364] How do we understand this situation? The Hopi description sounds like a freely swinging gyroscopic adjustment by the planet, making the carbon dioxide scenarios unnecessary. More interestingly, how did the Hopis know that the Earth was once frozen and motionless? Living in the high desert of Arizona, they can hardly have been expected to see ice as a major player in Earth history. Yet they have taught and believed the ice story for untold centuries, preceding Western scientists by a substantial amount of time.

The uniqueness of planet Earth. One more bit of astronomical data has emerged recently that will pose many problems for the religious side of the creation/evolution debate. Shankar Vedantan, writing in *The Denver Post* in August 2001, announced that "astronomers have found a planetary system remarkably similar to Earth's—two planets traveling in circular orbits around a star in the Big Dipper. The star is similar to the sun in chemical composition, and astronomers say the circular paths and sizes of the two planets hint at the presence of smaller, Earth-like bodies in tighter orbits." Vedantan went on to say that "more than 70 planetary systems have been discovered around stars other than our own, including three with multiple planets, but most have orbits that are sharply elliptical. Such orbits, which tend to freeze and heat the planet to extremes of temperatures as the planet dives close to a star and then pulls far away, are poor candidates for life to gain a foothold."[365]

We are on the brink of discovering that planet Earth is not unique in the cosmos. Many other planets probably have the necessary physical requirements for originating and sustaining life. They may also have had a far less catastrophic history than Earth, thereby enabling their biosphere to develop a full spectrum of life. Both Christianity and Western science put forward the proposition that Earth history represents a unique kind of cosmic reality; witness the confused logic of the anthropic principle, which assumes the universe has been designed so that people on Earth can understand it. Christianity arose when we had virtually no knowledge of the larger cosmos and this planet was presumed to be the center of creation. Christianity assumes that the life and death of Jesus is an event that affects the whole universe. The "last judgment" more or less wraps up the cosmic drama of atonement and redemption made necessary by the sin of the garden of Eden. Then a new heaven and Earth are created—falling into the almost archetypal beliefs of the non-Western peoples.

When we explore the heavens and meet creatures similar to ourselves from other planets, what will be our religious response to their wholly different perspective on the world? Although we picture aliens as monsters, there is no good reason to suppose they would be physically different than ourselves given the present premises of both science and religion. We would come to understand that our Earth history was a minor episode in the history of life in this region of space. We could be involved in interplanetary religious wars if we repeat the errors already committed in the discovery of the new world. The exclusive nature of Western religion, then, is applicable neither to Earth nor to other places with organic life in the universe.

Creation. We have not resolved the creation/evolution controversy, but we have certainly placed it in a different setting. Where does organic life come from? How does the Earth replenish its biosphere? We don't know, and our belief in Darwinism prevents us from considering alternative answers. What we do know is that at the beginning of every geologic era there seems to be a complete biotic system with prey-predator and symbiotic relationships. Jonathan Wells describes the actual situation we confront: "In Darwin's theory, the number of animal phyla gradually increases over time. The fossil record, however, shows that almost all of the animal phyla appear at about the same time in the Cambrian explosion, with the number declining slightly thereafter due to extinctions."[366] Wells quotes evolutionary theorist Jeffrey Schwartz as saying "the major animal groups 'appear in the fossil record as Athena did from the head of Zeus—full blown and raring to go.'"[367]

This fossil fact seems pretty much like a creation to me. We know that the scientific establishment will reject the idea out of hand. But so will the fundamentalists when they come to realize that god and/or the cosmic creative process continually produces a biosphere on planets that can support it. Boyce Rensberger, science columnist for the *Washington Post,* gives us the typical response of many scientists to the prospect of multiple creations. Mass extinctions are simply subsumed within the evolutionary folklore. "Mass extinction, then, sets the stage for many bursts of rapid evolution and the creation of entirely new forms of life."[368] If the loss of species sometimes runs as high as 90 percent, just how are rapid bursts of evolution going to make a fully developed biosphere, with more complex organisms, possible? Dead animals don't keep evolving, and if we accept Stephen Jay Gould's famous dictate, once species enter the fossil record, they have incredible stasis. We can only await another catastrophe to see the species disappear again. We will eventually be stuck with some version of creation. The *nth* term in this possible development would be scientists supporting and defending multiple creations and fundamentalists attacking the idea. Creation then will become an increasingly hot topic for discussion, and we can but hope that judges and justices learn more about recent scientific discoveries before they issue their judgments.

Protagonists' attitudes. We face an intellectual/emotional challenge of momentous proportions in the future. We can see only a very difficult road looming ahead until we resolve the parochial difficulties inside the Western theoretical framework and become oriented to the larger cosmic realities. The signposts of the future are very clear because we have so many scientists examining so many things, print and electronic media starving for stories they can sensationalize, and the vast majority of minds on both sides of the creation/evolution battle frozen and incapable of accepting new ideas about either science or religion. Scientists occupy the status of priests in our society, and they will not willingly surrender this favorable status—even if they have to lie to us. Fundamentalist preachers have somewhat the same attitude and cherish the idea that they are sources of wisdom with a direct line to god.

Michael Behe has an eloquent passage in *Darwin's Black Box* that summarizes what we can expect from the scientific establishment. "The history of science," he writes, "is replete with examples of basic-but-difficult questions being put on the back burner. For example, Newton declined to explain what caused gravity, Darwin offered no explanation for the origin or vision of life, Maxwell refused to specify a medium for light waves once the ether was debunked, and cosmologists in general have ignored the question of what caused the Big Bang."[369] We can anticipate, then, studied ignorance by the academic establishment and personal attacks on anyone who offers a new schema for arranging and interpreting the data we now have and will be receiving in the future. Even now, we have many instances of the establishment suppressing efforts to change orthodoxy. No one, it seems, wants to be considered different from his/her colleagues. W. B. Hamilton, writing in the *International Geology Review,* reported: "Peer review can represent the tyranny of the majority. I have run the peer-review gauntlet perhaps a hundred times. My papers describing and interpreting geology in more or less conventional terms have progressed smoothly, whereas publications of my manuscripts challenging accepted concepts have often been impeded, and occasionally blocked."[370] If we can foresee the need for substantial reform in geological thinking now that we have celestial impact events on the agenda, what will happen to people suggesting a greatly reduced geological timescale? What chance do we have to discuss the implications of the new data in the regular academic channels and publications?

The tendency in the Earth sciences will be to redouble efforts to defend the old faith. Claims will be made that radiometric dating devices are "greatly improved" and return reliable dates—dates coinciding with those that originated as a matter of speculation a century and a half ago. A. K. Baksi, writing in *Geology,* described the present use of the radiometric methods: "Subjective, and in many instances, incorrect use of radiometric data has become endemic in the earth science literature. Mathematical analysis of imperfect, and in many cases, highly subjective data sets lead to dubious conclusions."[371] When scientists pat us on the head and assure us that their dating is correct, we can but hope they have not tilted the scales to such a degree that we cannot rely on what they say.

Reforming science. How do we go about instituting reforms in academia so that new ideas can be heard? The academic environment is a strange kind of modern feudalism in which the senior scholars define the theoretical basis of the field—even if there is no good supporting evidence for their views. Moving down the academic ladder to full professors, associate and assistant professors, and graduate students, everyone is expected to adhere to the party line as defined by the clique dominant in the profession.

This discipline is something the Roman Catholic Church might envy because to be discredited in "the profession" means stagnation in career advancement, banishment from publication in prestigious journals, rejection of funding sources through peer review of proposals, and finally ostracism at professional meetings. Particularly when the subject is the doctrine of evolution, reform will be next to impossible for a long time. Avowing a belief in this nebulous concept is a must for scholars, comparable to affirming the Apostles' Creed in most of Christianity. As with Christianity, everyone has their favorite definition of evolution, and they project from a few stereotypical examples to affirm the universality of the doctrine. Evolution explains nothing, and it is not even necessary for most academic subjects except as a politically correct backdrop against which other theories are to be understood. Michael Behe noted that "a survey of thirty biochemical textbooks used in major universities over the past shows that many textbooks ignore evolution completely."[372]

Even more serious, however, is the possibility that what we are given in textbooks is misrepresented and may be deliberate fraud. Like ecclesiastics hiding some known fact about Jesus from their people, scientists may be deliberately presenting out-moded examples of evolution all the while knowing that what they are saying is not true. This possibility was first presented in a systematic fashion by Stephen Jay Gould in *The Mismeasure of Man,* in which he reviewed the work of his predecessors and showed how some had consciously altered evidence to fit their theories while others, subconsciously perhaps, had changed the selection of data to enable them to articulate their ideas. Instead of sparking a reform movement in science, however, the present situation even caught Gould in the web of deception.

Jonathan Wells devoted a major book, *Icons of Evolution,* to an examination of how certain "icons," that is to say, fundamental doctrines, were presented in a number of textbooks used in colleges and high schools. The major culprit seems to be the National Academy of Sciences (NAS), which issued a booklet in 1998 supporting the discredited Miller-Urey experiments with an alleged primordial biotic soup. In 1999 the NAS issued another booklet endorsing Darwin's "tree of life" that showed a continuous evolution of organisms and failed to mention the nature of the Cambrian explosion that contains most of the phyla we have ever seen. Also in 1999, the NAS released a booklet calling the supposed Galapagos finch evolution "compelling evidence" for Darwinism. Wells reviews the latest studies on Darwin's finches and shows that they have concluded that in drought years the finches with different-shaped beaks predominate, and in wet years they don't. They simply move back and forth between the two kinds of beaks. German evolutionist Ernst Haeckel's famous drawings of comparative embryos, reproduced in a large number of textbooks and touted as the real proof for Darwin's ideas, are shown to have been manipulated to prove his point, although they form one of the most familiar examples given to students to prove Darwinism. And the famous moths of England that were supposed to be modern examples of evolutionary adaptation do not actually behave as scientists tell us they do. The photos of them on tree trunks were faked.

Wells takes to task two of the prominent spokesmen for evolution, Douglas Futuyma and Stephen Jay Gould, on their support of Haeckel. Futuyma, it seems, authored a textbook on evolution for graduate students and simply included Haeckel's drawings without knowing that the theory had been thoroughly discredited long ago. Wells chastises Gould for having known about the faked Haeckel drawings for more than twenty years, failing to say anything about them, and then blaming their inclusion on "textbook writers," an apparently faceless, nameless group not representing science. If these people are the representatives of science, and they are certainly prominent in efforts to defend evolution, then the doctrine is in much worse shape than we have realized.

We thus come to the end of our inquiry. Reviewing the many topics we have discussed, we can see how subjects of such complexity could not easily be presented in an Arkansas or Louisiana courtroom. On the other hand, many scientists knew of the changes being made and understood that new studies would radically alter the way we look at the world. To support the old evolutionary scenario while understanding that it could not fit with the facts of modern science is an ethical problem of major proportions. We must ask when the scientific establishment will publicly admit that fundamental and irreversible change has occurred.

We also wonder why there is so much pressure on the various school boards to eliminate any criticism of evolution even when an opposing critique is necessary to understand the nature of the problem. Is the fear of criticism motivated by a distaste for creationists, or does it hide a deeper feeling of despair at the thought of surrendering a comfortable paradigm for an unknown future? In spite of the polls approving the teaching of creationism, no one of an educated mind wants Genesis taught in the classrooms as an alternative to science. Teaching bad science does not solve the problem either, and Jonathan Wells has exhaustively demonstrated that a substantial amount of misinformation and deliberate deceit is to be found in many evolutionary textbooks. Why are teachers who, in good faith, wish to give a broader explanation of the strengths and weaknesses of Darwinism, harassed and ridiculed? When is the American Civil Liberties Union going to catch up on its understanding of science?

My example of the Hopi memory of an icy planet illustrates the problem of reaching out to include the memories and insights of other cultures in the reconstruction of our knowledge of planetary history. Do we go ahead, as suggested by Ian Barbour, and construct a worldview that is wholly Western, and then, at the last moment, "find a place" for other cultures and their insights and beliefs? If almost every other society has believed that the planet is periodically destroyed by cosmic-size catastrophes, should not this possibility be thoroughly explored? Why must science cling to the idea of a linear time from the Big Bang to the present when the idea of linear time originates as a religious belief shared by a small portion of humanity? Obviously at the beginning of a new construction of Earth history, we must offer the most bizarre and creative scenarios for consideration if we are to solve the mystery of organic life on this planet. To cling to past paradigms and doctrines is not the way to proceed.

We have relied on authority figures in both science and religion, and they have brought us to this impasse. We are now expected to choose sides between two antagonists, neither of whom offers us an accurate and verifiable set of beliefs to follow. Neither Western religion nor Western science has an empirical foundation. Both apply outmoded doctrines to the data they do choose to examine. The Christian claim that god is working in history may be correct, but no one has made a decent case on behalf of this belief—ever. Organisms may well evolve, but they do not sneak offstage to do so. The anomalies in Western science and religion are so numerous that they now constitute an easily identifiable alternative to what we are presently asked to believe. We should demand that we be treated as adults—no more "Just So Stories" or religious myths need be fed to us.

ENDNOTES

Chapter One

1. S. Carpenter, "Kansas Cuts Evolution from Curriculum," *Science News* 156 (17 August 1999): 117.

2. Stephen Jay Gould, "Dorothy, It's Really Oz," *Time* (23 August 1999): 59.

3. Dennis Byrne, "Two Views of Same Reality," *Chicago Sun-Times*, 18 August 1999.

4. Hanna Rosin, "Creationism Evolves," *Washington Post*, 8 August 1999, National News section, A22.

5. Marjorie Coeyman, "Evolution Gets Dismissed from Some Classes," *Christian Science Monitor*, 16 August 1999.

6. Gordon Gregory, "College Instructor's Beliefs Endanger Job," *Seattle Times*, 20 February 2000, Local/Regional News section, B4.

7. Christine Haley, "Principal Sends Back Religious Books,"*AAAOL NEWS*, September 1999.

8. "Teaching Creationism," *Arizona Daily Star*, 14 March 2000, Editorial, 10A. See also James Glanz, "Poll Finds That Support Is Strong for Teaching 2 Origin Theories," *The New York Times*, 11 March 2000.

9. "Creationism Evolves," *Washington Post,* 8 August 1999, National News section, 22. See also Glanz, "Poll Finds That Support Is Strong."

10. Eric Eyre, "Group Abandons 'Creation' Textbook," *Charleston Gazette,* 4 April 2000.

11. Tim Talley, "Oklahoma House Passes Creationism Bill," Associated Press, 6 April 2000.

12. Julie Goodman, "Educators Discuss Evolution," Associated Press, 7 May 2000.

13. Paul Recer, "Teaching Evolution: States Get Graded," *The Denver Post,* 27 September 2000.

14. "Report Grades Evolution in States, Draws Fire from 'Intelligent Design' Advocates," 3 October 2000, www.atheists.org/flash.line/evol12.html.

15. John Milburn, "Kansas Restores Evolution to Science Classes," *Arizona Daily Star,* 15 February 2001, A8.

16. Francis X. Clines, "In Ohio School Hearing, a New Theory Will Seek a Place Alongside Evolution," *The New York Times,* 11 February 2002.

17. *Epperson v. Arkansas,* 393 U.S. 97 (1968).

18. *McLean v. Arkansas Board of Education,* 529 F. Supp. 1255 (1982).

19. *Edwards v. Aguillard,* 482 U.S. 578 (1987).

20. *Epperson v. Arkansas,* 113.

21. *McLean v. Arkansas Board of Education,* 1264.

22. Ibid.

23. Ibid., 1265.

24. Robert Cummings Neville, *Behind the Masks of God* (Albany: State Univ. of New York Press, 1991), 2.

25. Ibid.

26. Ibid., 15.

27. *McLean v. Arkansas Board of Education,* 1266.

28. *McLean v. Arkansas Board of Education,* 1268.

29. In *T. rex and the Crater of Doom* (Princeton: Princeton Univ. Press, 1997), Walter Alvarez advances the theory, based on the presence of a sediment containing rare minerals found all over the world, that a meteor hit the Yucatan area many millions of years ago, effectively eliminating the dinosaurs. His view has now become accepted and is regularly cited.

30. Some scholars are already hedging their bets in case uniformitarianism is wrong. Witness Robert Bakker's efforts in this respect: "A comet could follow a regular cycle of crashes with the earth, a trajectory of collisions repeated every time the comet's and the earth's orbits coincided. A mathematical analysis published in 1983 claimed that such extinctions struck regularly, every 26 million years, so the agent has even been dubbed with a name, the 26 Million Year Death Star." *The Dinosaur Heresies* (New York: William T. Morrow, 1986), 434. See also Boyce Rensberger, "Death of Dinosaurs: The True Story," *Newscience* 94, no. 8 (May 1986): 31. And these are the people who still deride Velikovsky!

31. *McLean v. Arkansas Board of Education*, 1269.

32. *Edwards v. Aguillard*, 589.

33. In 1970, in *Lemon v. Kurtzman*, 403 U.S. 602, 612–13, the Supreme Court articulated three considerations a state must meet in order to interfere with religious freedom via statutory enactment. They are: 1) a statute must have a secular purpose; 2) its principal or primary effect must be one that neither advances nor inhibits religion; and 3) the statute must not foster "an excessive governmental entanglement with religion."

34. *Edwards v. Aguillard*, 593–4.

35. Ibid., 630.

36. Ibid., 622.

37. Harold Booher, *Origins, Icons, and Illusions* (St. Louis: Warren H. Green, 1998), 314.

38. R. G. Collingwood, *The Idea of History* (New York: Galaxy Books, 1956), 255–56.

39. Robert Wright, "Science and Original Sin," *Time* (26 October 1996): 79.

40. Witness the comments by Albert Somit and Steven A. Peterson, eds., *The Dynamics of Evolution* (Ithaca, N.Y.: Cornell Univ. Press, 1989), 2: "Some of

those who led the opposition to punctuationism made no secret of their distaste for its potential social and political overtones. For instance, some have noted that punctuational theory may provide at least metaphoric support for advocacy of revolutionary change."

41. Ian Barbour notes in *Religion in an Age of Science* (San Francisco: HarperSanFrancisco, 1990), 106–7: "In 1958 John Bell calculated the statistical correlation one would expect between the two detectors (as a function of their relative orientations) if Einstein's assumptions are correct. Recent experiments by Alain Aspect and others (using photons rather than protons) have not been consistent with these expectations, indicating that one of Einstein's assumptions is incorrect."

42. This strange interpretation of evolution suggests that evolution is not a steady movement of change but rather a series of sporadic jumps in organic development. Allegedly, this theory is built on fossil evidence that shows completely developed organisms during any of the geologic eras, generally at the beginning of them. Logically there is no way that this data can prove evolution, since all that can be said is that we find a rather impressive biotic system in each geologic era.

43. Niles Eldredge, *Time Frames* (New York: Simon and Schuster, 1985), 15.

44. Bakker, *Dinosaur Heresies,* 401.

45. Eldredge, *Time Frames,* 141.

46. Ibid., 144.

47. In 1950 Immanuel Velikovsky published *Worlds in Collision* (New York: Macmillan), in which he suggested that at some point in ancient history, Mars and Venus had come into near collision with the earth. In 1958, at the start of the International Geophysical Year, he offered several challenges that, he argued, would verify his scenario. Many of his predictions were proven accurate, the most spectacular being the idea that the earth has an electrical field surrounding it. The Van Allen Belts were found to support Velikovsky.

48. Already some scientists are braving the scorn of their colleagues and suggesting that they have found evidence of Noah's flood. See "Deep-Sea Explorers Say Find Points to Noah's Flood," *Tampa Tribune,* 25 November 1999. See also "Comet That Launched Noah's Ark," 1996, Nado.net.

49. *Tangipahoa Parish Board of Education v. Frelier,* 147 L.Ed. 2d. 974, 68 USLW 3657, 68 USLW 3770, 68 USLW 3771.

Chapter Two

50. Hans Küng, *Theology for the Third Millennium* (New York: Doubleday, 1988), 134.

51. *McLean v. Arkansas Board of Education,* 1267.

52. Pierre Teilhard de Chardin, *The Phenomenon of Man* (New York: Harper Torchbooks, 1961), 219.

53. Paul Davies, *The Mind of God* (New York: Simon and Schuster, 1992), 84.

54. Keith Ward, *God, Faith, and the New Millennium* (Oxford: One World Publications, 1998), 52.

55. Richard Milton, *Forbidden Science* (London: Fourth Estate, 1994), 218.

56. Hans Schaer, *Religion and the Cure of Souls in Jungian Psychology* (New York: Bollingen Series XXI, Pantheon Books, 1950), 55.

57. Ibid.

58. Langdon Gilkey, *Creationism on Trial* (Minneapolis: Winston Press, 1985), 180.

59. Norman F. Hall and Lucia K. B. Hall, "Is the War between Science and Religion Over?" *The Humanist* (May/June 1986): 26.

60. Ward, *God, Faith, and the New Millennium,* 66.

61. Steven Stanley, *The New Evolutionary Timetable* (New York: Basic Books, 1981), 174.

62. Robert Lowie, *Primitive Religion* (New York: Liveright, 1948), 342.

63. Booher, *Origins, Icons, and Illusions,* 355.

64. Ibid., 41.

65. Phillip Johnson, *Darwin on Trial* (Washington, D.C.: Regnery Gateway, 1991), 151.

66. Stephen Jay Gould, "Punctuated Equilibrium in Fact and Theory," in Somit and Peterson, *Dynamics of Evolution*, 73.

67. Gilkey, *Creationism on Trial*, 150–51.

68. Michael Behe, *Darwin's Black Box* (New York: Free Press, 1986), 185–86.

69. Ibid., 72.

70. Ibid., 176.

71. Philip Kitcher, *Abusing Science* (Cambridge, Mass.: MIT Press, 1982), 71.

72. Ian Barbour, *Religion and Science* (San Francisco: HarperSanFrancisco, 1997), 94.

Chapter Three

73. Paul Feyerabend, *Science in a Free Society* (New York: LNB, 1978), 74.

74. Paul Feyerabend, *Against Method* (New York: Schocken Books, 1975), 216.

75. Stephen Jay Gould, *Rocks of Ages* (New York: The Ballantine Publishing Groups, 1999), 213.

76. Ibid., 21–3.

77. Kenneth Woodward, "Is God Listening?" *Newsweek* (31 March 1997): 59.

78. Gilkey, *Creationism on Trial*, 177.

79. Davies, *Mind of God*, 58.

80. Feyerabend, *Against Method*, 299.

81. Barbour, *Religion and Science* (San Francisco: HarperSanFrancisco, 1997), 126.

82. Feyerabend, *Science in a Free Society*, 88.

83. Percy W. Bridgman, *The Way Things Are* (Cambridge, Mass.: Harvard Univ. Press, 1959), 129.

84. J. Richard Greenwell, "The Dinosaur Vote," *Science Digest* 90, no. 4 (April 1992): 42.

85. Ibid.

86. Preston Douglas, "The Lost Man," *The New Yorker* (1998): 77.

87. W. D. Rouse, *Catalysts for Change* (New York: John Wiley & Sons, 1993), 81–2. Rouse is a systems engineer in Atlanta, Georgia, who writes on the management problems of Fortune 500 companies. He has found the same conceptual difficulties in science as in private business. Quoted in Booher, *Origins, Icons, and Illusions,* 351.

88. J. V. McConnell, "Science and Pseudoscience," *Skeptical Inquirer* 11, no. 1: 104–5.

89. Thomas Kuhn, *The Structure of Scientific Revolutions* (Chicago: Univ. of Chicago Press, 1970), 37.

90. Richard Milton, *Shattering the Myths of Darwinism* (Rochester, Vt.: Park Street Press, 1997), 50–1.

91. J. Ogden, "The Use and Abuse of Radiocarbon," *Annals of the New York Academy of Science* 288 (1977): 173.

92. Barbour, *Religion and Science,* 53.

93. Davies, *Mind of God,* 23.

94. Philip Slater, *The Wayward Gate* (Boston: Beacon Press, 1977), 67.

95. Kenneth Ring, *Life at Death* (New York: Coward, McCann and Geoghegan, 1980), 219.

96. Kuhn, *The Structure of Scientific Revolutions,* 122–23.

97. Ibid., 19. As a humorous aside, Stephen Jay Gould remarked in *Ever Since Darwin* (New York: W. W. Norton & Co., 1977), 167, that "I am not distressed by the crusading zeal of plate tectonics, for two reasons. My intuition, culturally bound to be sure, tells me it is basically true." Do we have a group intuition here or merely a recognition of the power of peer opinion?

98. Kuhn, *The Structure of Scientific Revolutions,* 90.

99. Feyerabend, *Science in a Free Society,* 88.

100. David Foster, *The Philosophical Scientists* (New York: Dorset Press, 1985), 147.

101. Edward T. Hall, *Beyond Culture* (New York: Anchor-Doubleday, 1976), 151.

102. Foster, *Philosophical Scientists,* 150.

103. Rupert Sheldrake, "Is the Universe Alive?" *The Teilhard Review* 25, no. 1

(Spring 1990): 16.

104. F. David Peat, *Synchronicity* (New York: Bantam Books, 1987), 97.

105. Werner Heisenberg, *Across the Frontiers* (New York: Harper and Row, 1976), 111.

106. There has been much recent publicity about the speed of light being super-seded in an experiment, and a new genre of scientific literature appears to be gathering about this topic. See Ricardo Carezani, "A General Discussion on Limit Velocity," www.autodynamics.org/new99/Atomic/Limit Velocity/index.html; William J. Cromie, "Physicists Slow Speed of Light," www.news.harvard.edu/science/current_stories/18.Feb.99/light.html; Philip Gibbs, "Is the Speed of Light Constant?" www.math.ucr.edu/home/baez/physics/speed_of_light.html. These papers illustrate how tentative our "absolute" concepts really are.

107. Heisenberg, *Across the Frontiers*, 22. Heisenberg elaborated on this idea on page 116: "I think on this point modern physics has definitely decided for Plato. For the smallest units of matter are in fact not physical objects in the ordinary sense of the word: they are forms, structures, or—in Plato's sense—Ideas, which can be unambiguously spoken of only in the language of mathematics."

108. Ibid., 16.

109. Rupert Sheldrake, *The Rebirth of Nature* (New York: Bantam Books, 1980), 43.

110. Percy W. Bridgman, *The Logic of Modern Physics* (New York: Macmillan, 1946), 22.

111. Werner Heisenberg, *The Physicist's Conception of Nature* (New York: Harcourt, Brace and Co., 1955), 28–9.

112. Werner Heisenberg, *Physics and Philosophy* (New York: Harper Torchbooks, 1958), 58.

113. Fred Alan Wolf, *The Dreaming Universe* (New York: Simon and Schuster, 1994), 162.

114. Heinz Pagels, *Perfect Symmetry* (New York: Bantam Books, 1986), 67.

115. Booher, *Origins, Icons, and Illusions*, 153.

116. Charles W. Petit, "From Big Bang to Big Bounce," *U.S. News and World*

Report (6 May 2002): 59.

117. Davies, *Mind of God,* 86.

118. David Lindley, *The End of Physics* (New York: Basic Books, 1993), 205–6, quoted in Booher, *Origins, Icons, and Illusions,* 153.

119. Sheldrake, *Rebirth of Nature,* 57.

Chapter Four

120. Milton, *Shattering the Myths of Darwinism,* 130.

121. D. E. Tyler, *Originations of Life* (Ontario, Ore.: Discovery Books, 1983), 77.

122. George Gaylord Simpson, *The Major Features of Evolution* (New York: Simon and Schuster, 1953), 164.

123. Ibid., 168.

124. Ibid., 342.

125. Ibid., 175.

126. Ibid., 30.

127. Bakker, *Dinosaur Heresies,* 449.

128. A recent article by Susan Milius, "Alarming Butterflies and Go-Getter Fish," *Science News* 160 (21 July 2001): 42, discusses the new conceptions of species. She quotes Dolph Schluter of the University of British Columbia as "one who is taking a new look at how species form." When asked for a rough working definition of species, he gives a minor variation on the familiar theme: Members of a species share some distinctive look or makeup and don't breed a lot with outsiders. The same thing could be said of the societies at Ivy League colleges or the typical terrorist group.

129. Stephen Jay Gould, *The Mismeasure of Man* (New York: W. W. Norton & Co., 1981), 44.

130. Ibid.

131. Matthew A. Bille, *Rumors of Existence* (Blaine, Wash.: Hancock House, 1995), 14.

132. Gordon Rattray Taylor, *The Great Evolution Mystery* (New York: Harper &

Row, 1983), 143–44.

133. Simpson, *Major Features of Evolution*, 77.

134. Ibid., 256–57.

135. Ibid., 63.

136. Stanley, *New Evolutionary Timetable*, 133.

137. Robert Wesson, *Beyond Natural Selection* (Cambridge, Mass.: MIT Press, 1993), 209–10.

138. See David Hurst Thomas, *Skull Wars* (New York: Basic Books, 2000), and Roger Downey, *The Riddle of the Bones* (New York: Springer-Verlag New York, 2000).

139. Eldredge, *Time Frames*, 113.

140. Sir Fred Hoyle and N. Chandra Wickramasinghe, *Evolution from Space* (New York: Simon and Schuster, 1981), 127.

141. Stanley, *New Evolutionary Timetable*, 83–84.

142. Wesson, *Beyond Natural Selection*, 207.

143. Cited in Booher, *Origins, Icons, and Illusions*, 62, summarizing Boyce Rensberger, "Science," *Washington Post*, 19 July 1993, A3.

144. Steven Stanley, "The Empirical Case for the Punctuated Model of Evolution," in Somit and Peterson, *Dynamics of Evolution*, 93.

145. Eldredge, *Time Frames*, 81.

146. Booher, *Origins, Icons, and Illusions*, 379.

147. Norman Macbeth, *Darwin Retried* (Cambridge, Mass.: Harvard Common Press, 1979), 37.

148. Kitcher, *Abusing Science*, 119.

149. Adrian Desmond, *Hot-Blooded Dinosaurs* (New York: The Dial Press/James Wade, 1976), 156.

150. Ibid.

151. Stanley, *New Evolutionary Timetable*, 93.

152. Ibid., 86.

153. Bakker, *Dinosaur Heresies,* 139.

154. Ibid., 70.

155. Ibid., 194–95.

156. Ibid., 172.

157. Stanley, *New Evolutionary Timetable,* 129.

158. Desmond, *Hot-Blooded Dinosaurs,* 96.

159. Ward, *God, Faith, and the New Millennium,* 116.

160. Ibid., 118.

161. Stephen Jay Gould, *Science Digest* (May 1986): 32, cited in Erich von Fange, *Noah to Abraham* (Syracuse, Ind.: Living Word Services, 1994), 34.

162. Eldredge, *Time Frames,* 105.

Chapter Five

163. Ibid.

164. Derek Ager, *The Nature of the Stratigraphic Record* (New York: John Wiley & Sons, 1981), 68. Even the *Encyclopedia Britannica* rejects this train of logic: "It cannot be denied that from a strictly philosophical standpoint geologists are here arguing in a circle. The succession of organisms has been determined by a study of their remains embedded in the rocks, and the relative ages of the rocks are determined by the organisms that they contain." R. H. Rastal, vol. X, 168.

165. Alfred de Grazia, *The Lately Tortured Earth* (Princeton: Metron Publications, 1983), 360.

166. E. M. Durrance, *Radioactivity in Geology* (Chister, England: Ellis Horwood Ltd., 1986), 287.

167. Ager, *Nature of the Stratigraphic Record,* 27.

168. Ibid., 49. Ager further stated: "It has been calculated that, in the Gulf of Mexico, there is a 95% probability that a hurricane will pass over a particular point on the coast at least once in 3000 years. The maximum amount of sediment likely to be deposited over that period along the coast generally is

about 30 cm and we know that hurricanes will certainly rearrange that amount of material"(75).

169. Derek Ager, *The New Catastrophism* (Cambridge, Mass.: Cambridge Univ. Press, 1993), 78.

170. De Grazia, *Lately Tortured Earth*, 328.

171. Dolph Earl Hooker, *Those Astounding Ice Ages* (New York: Exposition Press, 1958), 127.

172. Robert S. Boyd, "Ocean Floor Baring Secrets," *The Denver Post*, 27 August 2000, 41A.

173. Hooker, *Those Astounding Ice Ages*, 104.

174. Ager, *Nature of the Stratigraphic Record*, 7.

175. Ibid., 9.

176. Ibid., 2.

177. Ibid., 6.

178. Ibid., 8.

179. Ibid., 13.

180. Here the fundamentalists score many points. According to John G. Whitcomb Jr.and Henry M. Morris, *The Genesis Flood* (Philadelphia: Prespyterian and Reformed Publishing Co., 1964), 163: "The Dismal Swamp of Virginia, perhaps the most frequently cited case of a potential coal bed, has formed an average of 7 feet of peat, hardly enough to make a single respectable seam of coal. Furthermore, there is no actual evidence that peat is now being transformed into coal anywhere in the world. No locality is known where the peat bed, in its lower reaches, grades into a typical coal bed."

181. Immanuel Velikovsky, *Earth in Upheaval* (Garden City, N.Y.: Doubleday, 1955), 217.

182. Hooker, *Those Astounding Ice Ages*, 128–29.

183. "Comet Shoemaker-Levy Background," www.jpl.nasa/gov/s19/background.html, June 20, 2002.

184. Derek Allan and J. Bernard Delair, *When the Earth Nearly Died* (Bath, England: Gateway Books, 1995), 200. Actually, Ignatius Donnelly raised the

question of Lexell's comet in Ragnarok more than a century ago (85).

185. David Kring, "Impact Events and Their Effect on the Origin, Evolution, and Distribution of Life," *GSA Today* 10, no. 8 (August 2000): 2.

186. Alvarez, *T. rex and the Crater of Doom*, 140.

187. Victor Clube and Bill Napier, *The Cosmic Serpent* (New York: Universe Press, 1982), 96.

188. Ager, *Nature of the Stratigraphic Record*, 34.

189. Robert H. Dott, "The Rule," Presidential Address to the Society of Economic Paleontologists and Mineralogists, *Geotimes* (November 1982): 16. This change extends even to some scientific reporters, who are now hedging their bets. Witness Rensberger, "Death of Dinosaurs: The True Story?": "More significant still is the growing conviction among scientists that cataclysmic mass extinctions, perhaps brought on by similar events, are recurrent phenomena that have been playing a far more important role in the course of evolution than almost anyone had hitherto suspected" (31).

190. Ager, *Nature of the Stratigraphic Record*, 40–1.

191. Hooker, *Those Astounding Ice Ages*, 85.

192. Ibid.

193. Ibid., 80.

194. Ager, *Nature of the Stratigraphic Record*, 210.

195. Brian J. Skinner and Stephen C. Porter, *The Dynamic Earth* (New York: John Wiley & Sons, 1992), 450–64.

196. Donald Patten, *The Biblical Flood and the Ice Epoch* (Seattle: Pacific Meridian Publishing, 1966), 78–9.

197. Donald Patten, Ronald R. Hatch, and Loren Steinhaur, *The Long Day of Joshua and Six Other Catastrophes* (Seattle: Pacific Meridian Publishing, 1973), 308 (footnote).

198. Alfred de Grazia, *Chaos and Creation* (Princeton, N.J.: Metron Publications, 1981), 45.

199. Ager, *New Catastrophism*, 163.

Chapter Six

200. Ibid., 162.

201. Clube and Napier, *Cosmic Serpent*, 139.

202. Paul Recer, "Experts: Asteroid Won't Hit Earth,"Associated Press, 3 March 1998.

203. Geri Destefano and Alfred Webre, "Comet 'Lee' Possible Connection with CME's," *EcoNews Service;* email, econews@ecologynew.com.

204. NASA press release, July 22, 1999.

205. Ian Barbour, *Issues in Science and Religion* (New York: Harper Torchbooks, 1966), 7.

206. Frank Tipler, *The Physics of Immortality* (New York: Doubleday, 1994), 3.

207. Langdon Gilkey, "Social and Intellectual Sources of Contemporary Protestant Theology in America," in eds., *Daedalus: Religion in America,* winter 1967, 71.

208. Barbour, *Religion and Science,* 331.

209. Arnold Toynbee, *An Historian's Approach to Religion* (London: Oxford Univ. Press, 1956), 141.

210. Murray Wax, "Religion As Universal: Tribulations of an Anthropological Enterprise," *Zygon* 19, no. 1 (March 1984): 6.

211. Ibid.

212. Ward, *God, Faith, and the New Millennium,* 141.

213. Barbour, *Religion and Science,* 87.

214. A. McG. Beede, *Western Sioux Cosmology* and *Letting Go of the Ghost* (Chicago: Newberry Library, n.d.), 6.

215. Ibid., 7.

216. Ibid., 9.

217. Küng, *Theology for the Third Millennium,* 223.

218. Robert Bellah, *Beyond Belief* (New York: Harper and Row, 1970), 42.

219. G. Van Der Leeuw, *Religion in Essence and Manifestation* (London: George

Allen & Unwin, 1938), 53.

220. Barbour, *Issues in Science and Religion,* 415.

221. Ibid., 5.

222. Barbour, *Religion and Science,* 97.

223. Ward, *God, Faith, and the New Millennium,* 175.

224. Sharon Begley and Marian Wesley, "Science Finds God," *Newsweek* (20 July 1998): 50.

225. Van Der Leeuw, *Religion in Essence and Manifestation,* 79.

226. Ibid., 159.

227. Eric J. Sharpe, *Nathan Söderblom and the Study of Religion* (Chapel Hill, N.C.: Univ. of North Carolina Press, 1990), 169–70.

228. Ward, *God, Faith, and the New Millennium,* 154.

229. Hans Küng, *Eternal Life* (New York: Doubleday, 1984), 48–9.

230. Ernest Benz, "On Understanding Non-Christian Religions," in Louis Schneider, ed., *Religion, Culture, and Society* (New York: John Wiley & Sons, n.d.), 5.

231. Ibid.

232. Gilkey, *Creationism on Trial,* 217.

233. Ian Barbour, *Myths, Models, and Paradigms* (San Francisco: Harper and Row, 1974), 23.

234. Küng, *Theology for the Third Millennium,* 244.

235. Arnold Toynbee, *A Study of History* (New York: Oxford Univ. Press, 1947), 39.

236. Teilhard de Chardin writes, in *The Phenomenon of Man* (New York: Harper Torchbooks, 1961), 212: "We would be allowing sentiment to falsify the facts if we failed to recognize that during historic times the principal axis of anthropogenesis has passed through the West. It is in this ardent zone of growth and universal recasting that all that goes to make man today has been discovered, or at any rate *must have been discovered*. For even that which had long been known elsewhere only took on its definitive human value in becoming incorporated in the system of European ideas and activities."

237. Küng, *Theology for the Third Millennium*, 244.

238. Küng, *Eternal Life*, 132.

Chapter Seven

239. Paul Tillich, *Christianity and the Encounter of the World Religions* (New York: Columbia Univ. Press, 1963), 65.

240. Sheldrake, "Is the Universe Alive?" 19.

241. Tillich, *Christianity and the Encounter of the World Religions*, 17.

242. Barbour, *Religion and Science*, 189.

243. Foster, *Philosophical Scientists*, 9.

244. Gilkey, *Creationism on Trial*, 107.

245. Ward, *God, Faith, and the New Millennium*, 154.

246. Paul Tillich, *The Protestant Era* (Chicago: Univ. of Chicago Press, 1948), 99–100.

247. Van Der Leeuw, *Religion in Essence and Manifestation*, 27.

248. Ibid., 27.

249. H. R. Hays, *In the Beginning* (New York: Putnam, 1963), 188.

250. Ibid., 424.

251. Peat, *Synchronicity*, 123.

252. Hays, *In the Beginning*, 475.

253. Wolf, *Dreaming Universe*, 206.

254. Benz, "On Understanding Non-Christian Religions," 4.

255. Nathan Söderblom, *The Living God* (London: Oxford Univ. Press, 1933), 315.

256. Van Der Leeuw, *Religion in Essence and Manifestation*, 185.

257. Ibid., 29–30.

258. Hays, *In the Beginning*, 533.

259. Davies, *Mind of God*, 43.

260. Ibid.

261. Barbour, *Religion and Science*, 58.

262. Bellah, *Beyond Belief*, 28.

263. Ibid., 22.

264. Ibid.

265. Söderblom, *The Living God*, 231–32.

266. Huston Smith, *The Religions of Man* (New York: Harper and Row, 1958), 164.

267. Bellah, *Beyond Belief*, 33.

268. Tillich, *Protestant Era*, 120.

269. Ibid., 123.

270. Robert Bellah, *Tokugawa Religion* (Glencoe, Ill.: The Free Press, 1957), 62.

271. James Gustafson, "Introduction," in H. Richard Niebuhr, *The Responsible Self* (New York: Harper & Row, 1963), 14.

272. See, for example, Barbour, *Myths, Models, and Paradigms*, 178: "For the Christian, this path involves the recognition that God has been at work in other religious traditions; their faith and thought may be genuine responses to God in the context of their cultural assumptions." See also John Macquarrie, *Three Issues in Ethics* (New York: Harper & Row, 1970), 138: "There are people who do not believe in any power making for righteousness in this world, who have no explicit hope that the world process is directed toward any meaningful end, who, to put it briefly, do not believe in God in any sense, and yet are not victims of the kind of moral paralysis which we have envisaged. Such persons are sometimes wholeheartedly devoted to moral and social ideals, and their ideals may be very close to those of the Christian religion." See also Niebuhr, *Responsible Self*, 144: "And on the other hand we do not fail to note that among our companions who refuse to take the name of Christian responses to action are many that seem to be informed by the trust, the love of all being, the hope in the open future, that have become possible to us only in our life with Jesus Christ. ..."

273. Barbour, *Issues in Science and Religion*, 20.

Chapter Eight

274. The most radical early dating is done by Zecharia Sitchin in *The Twelfth Planet* (New York: Stein and Day, 1976), 410, in which he projects a date of 445,000 years ago as when the Nephilim arrived on earth. I cite his estimate to indicate that considerable discrepancy exists between biblical ages and ages that can be obtained from other ancient sources.

275. Skinner and Porter, *Dynamic Earth*, 169.

276. Richard Mooney, *Gods of the Air and Darkness* (New York: Stein and Day, 1975), 130.

277. Ibid., 132.

278. Giorgio Santillana and Hertha von Dechend, *Hamlet's Mill* (Boston: Gambit Books, 1969), 119.

279. Ibid.

280. Patten, *Biblical Flood*, 167.

281. Sitchin, *Twelfth Planet*, 189–91.

282. The discovery in the Swiss Alps of a man from ancient times completely surprised scholars who had supported the "barely savage" stereotype of human accomplishments at the end of the Ice Age, and they quickly back-tracked and pretended they had believed in an advanced civilization from the very beginning.

283. Peter Tompkins, *Mysteries of the Mexican Pyramids* (New York: Stein and Day, 1975), 350–51.

284. Louis Ginzberg, *The Legends of the Jews,* vol. 1 (Philadelphia: Jewish Publishing Society of America, 1925), 4.

285. Immanuel Velikovsky, "The Hebrew Cosmology," *In the Beginning,* www.varchive.org/itb/index.html.

286. Sitchin, *Twelfth Planet,* 53.

287. I. S. Shklovskii and Carl Sagan, *Intelligent Life in the Universe* (San Francisco: Holden-Day, 1966), 459.

288. Jim Bailey, in *The God-Kings and the Titans* (New York: St. Martin's Press, 1973), 181–82, cited Herodatus I: "These same Chaldeans say (but I do not believe them) that the god himself is wont to visit the shrine and rest upon

the couch, even as in Thebes of Egypt, as the Egyptians say (for there too a woman sleeps in the temple of the Theban Zeus, and neither the Egyptian nor the Babylonian woman, it is said, has intercourse with men) as it is likewise with the prophetess of the god at Patara in Lycia." Sounds like astronaut hanky-panky to me.

289. Barbour, in *Religion and Science*, 200, reports a date of the fifth century B.C.E., which would provide plenty of time to lose most of the data that would provide an accurate chronology.

290. Ibid., 454.

291. Sitchin, *Twelfth Planet*, 302.

292. Barbour, *Religion and Science*, 200.

293. Barbour, *Myths, Models, and Paradigms*, 154–55.

294. Ibid.

295. Marvin Pope, "EL in the Ugaritic Texts," in Leiden, ed., *Supplement to the Vetus Testamentum*, vol. II (Leiden, Netherlands: E. J. Brill, 1955), 18–19.

296. Ibid., 29.

Chapter Nine

297. James Glanz, "Intelligent Design Grows as Challenge to Evolution," *The Denver Post*, 8 April 2001.

298. Victor J. Stenger, "Intelligent Design," *Talk Origins Archives*, 8 June 2000.

299. Ibid.

300. www.reasons.org/resources/papers/design.html, pp. 8–10

301. William J. Broad, "Apollo Opened Windows on Moon's Violent Birth," *The New York Times*, 20 July 1999, Science Times section, D2.

302. Foster, *Philosophical Scientists*, viii.

303. Stenger, "Intelligent Design."

304. John Noble Wilford, "New Findings Help Balance Cosmological Books," *The New York Times*, 9 February 1999, Science Times section, D9.

305. John D. Barrow and Frank J. Tipler, *The Anthropic Cosmological Principle*

(New York: Oxford Univ. Press, 1986), 21.

306. Ibid., 23.

307. Ibid.

308. Barbour, *Religion and Science*, 205.

309. Behe, *Darwin's Black Box*, 247.

310. Hugh Ross, "Design and the Anthropic Principle," www.reasons.org/resources/papers/design.html.

311. Wilford, "New Findings Help Balance Cosmological Books," D9.

312. Margaret Wertheim, "The Odd Couple," *The Sciences* (March/April 1999): 39.

313. James Glanz and Dennis Overbye, "Universe's Constants Could Prove Otherwise," *The Denver Post*, 15 August 2001, 2A.

314. Wertheim, "The Odd Couple," 40.

315. Scholarly papers on this subject seem to be proliferating rapidly. See: Michael Fowler, "The Speed of Light," www.phys.Virginia.edu/classes109N/lectures/spedlite.html; Gibbs, "Is the Speed of Light Constant?"; Cromie, "Physicists Slow Speed of Light"; Carezani, "A General Discussion on Limit Velocity."

316. John R. Hadd, *Evolution: Reconciling the Controversy* (Glassboro, N.J.: Kronos Press, 1979), 55.

317. Ibid., 60.

318. Booher, *Origins, Icons, and Illusions*, 305.

319. Foster, *Philosophical Scientists*, 165–66.

320. Booher, *Origins, Icons, and Illusions*, 160.

321. Foster, *Philosophical Scientists*, 169.

322. William A. Demski, "Intelligent Design as a Theory of Information," www.arn.org/docs/demski/wd_idtheory.html.

323. Sheldrake, *Rebirth of Nature*, 57.

324. Barbour, *Religion and Science*, 190.

325. John Yaukey, "Animal Intellect Gaining Respect," *The Denver Post,* 19 September 1999, 6A.

326. Ibid., 2.

327. Dimitia Smith, "Can Animals Think? Parrot Seems to Say Yes," *The Denver Post,* 16 October 1999, 32A.

328. Yaukey, "Animal Intellect Gaining Respect," 6A.

329. Charles Eastman, *Indian Boyhood* (New York: McClure, Phillips & Co, 1902), 160–61.

330. Barbour, *Religion and Science,* 287.

331. Frances Densmore, "Teton Sioux Music," Bureau of American Ethnography, Bulletin 61, Smithsonian Institution (1918): 205.

332. Ibid., 210.

Chapter Ten

333. John Noble Wilford, "From Pluto and Saturn, Clues to Solar System's Birth," *The New York Times,* 10 August 1999, p. x.

334. Decades ago I read about this possibility in George Gamow's *One, Two Three, Infinity* (New York: Viking Press, 1947) and thought then it was a splendid idea.

335. Broad, "Apollo Opened Windows on Moon's Violent Birth," D2.

336. Lee Bowman, "Scientists Rethink Size of Collisions with Earth," *Minneapolis Star-Tribune,* 3 August 1997, 4A.

337. Allan and Delair, *When the Earth Nearly Died,* 214–15.

338. Robert S. Boyd, "Craters Hunters Finding Multiple-Impact Evidence," *The Denver Post,* 4 October 1998, 26A.

339. Douglas Isbell and Mary Hardin, "Chain of Impact Craters Suggested by Spaceborne Radar Images," *Comet Shoemaker-Levy Home Page,* 29 March 1996.

340. James Trefil, "Ancient Armageddon," *USA Weekend,* 26–28 June 1998, 10.

341. Boyd, "Craters Hunters Finding Multiple-Impact Evidence," 26A.

342. Ibid.

343. U.S. Geological Survey, "Comet Struck Southern Nevada 370 Million Years Ago," News Release, 20 October 1997.

344. Christina B. Claussen, "Meteor Impact in the Barents Sea," Foundation of Scientific and Industrial Research of the Norwegian Institute of Technology, Trondheim, Norway, January 1999.

345. Greg Smith, "Iowa Saw 'Ancient' Armageddon," *The Denver Post,* 5 July 1998, 4A.

346. William J. Broad, "Source of Die-Off Found?" *Denver Post,* 30 April 2000, 34A.

347. Ignatius Donnelly, *Ragnarok: The Age of Fire and Gravel* (New York: University Books, 1970), 85.

348. Allan and Delair, *When the Earth Nearly Died,* 200.

349. Kuhn, *The Structure of Scientific Revolutions,* 116.

350. Alvarez, *T. rex and the Crater of Doom,* 140.

351. Robert S. Boyd, "Cosmic Collisions Shaped Life on Earth," *The Denver Post,* 22 August 1999, 27A.

352. Rensberger, "Death of Dinosaurs: The True Story," 78.

353. J. Madeleine Nash, "When Life Nearly Died," *Time* 146, no. 12 (18 September 1995), www.time.com.mag...earchive/1995/950918/9500.18.science.html. Scientists give the impression that only animal-type creatures died at the end of the Permian, but Maggie Fox, in "Big Extinctions Wiped Out Plants, Too, Study Finds," Reuters, 8 September 2000, noted: "Evidence from South Africa's Karoo region suggests strongly that many species of plants disappeared at the end of the Permian period, along with oceans teeming with creatures."

354. Dale A. Russell, "The Enigma of the Extinction of the Dinosaurs," *Annual Review of Earth and Planetary Sciences* 7 (1979): 176.

355. Alan R. Hildebrand, "The Cretaceous/Tertiary Boundary Impact (Or: The Dinosaurs Didn't Have a Chance)," *Journal of the Royal Astronomical Society of Canada* 87, no. 2 (1993): 111.

356. S. Perkins, "Was It Sudden Death for the Permian Period?" *Science News* 158 (15 July 2000): 39.

357. "Space Impact Suggested by New Evidence on Extinctions," *Arizona Daily Star*, 11 May 2001, A5.

358. Kenneth Chang, "Meteor May Have Started Dinosaur Era," *The New York Times*, 17 May 2002.

359. Boyd, "Cosmic Collisions Shaped Life on Earth," 27A.

360. Richard Monastersky, "Popsicle Planet," *Science News* 154 (29 August 1998): 139.

361. R. H., "When Glaciers Covered the Earth," *Science News* 151 (29 March 1997): 196.

362. Ibid.

363. Frank Waters, *Book of the Hopi* (New York: Viking Press, 1963), 16.

364. Kenneth Chang, "Scientists Try to Explain the Cold, Mysterious Era of 'Snowball Earth,'" *The New York Times*, 19 June 2001.

365. Shankar Vedantan, "Solar System Like Earth's May Point to Planets with Water," *The Denver Post*, 16 August 2001, 8A. The increasing popularity of this idea was further demonstrated by David H. Levy, "The Search for Other Worlds," *Parade* (30 September 2001): 4–6.

366. Wells, *Icons of Evolution*, 43.

367. Ibid., 41.

368. Rensberger, "Death of Dinosaurs: The True Story," 32.

369. Behe, *Darwin's Black Box*, 251.

370. W. B. Hamilton, "Archean Tectonics and Magnetism," *International Geology Review* 40 (1998): 3.

371. A. K. Baksi, "Search for Periodicity in Global Events in the Geological Record: Quo Vadimus?" *Geology* 18 (1990): 985.

372. Behe, *Darwin's Black Box*, 182.

BIBLIOGRAPHY

Ager, Derek. *The Nature of the Stratigraphic Record*. New York: John Wiley & Sons, 1973.

———. *The New Catastrophism*. Cambridge, Mass.: Cambridge Univ. Press, 1993.

Allan, D. S., and J. B. Delair. *When the Earth Nearly Died*. Bath, England: Gateway Books, 1995.

Alvarez, Walter. *T. rex and the Crater of Doom*. Princeton, N.J.: Princeton Univ. Press, 1997.

Bailey, Jim. *The God-Kings and the Titans*. New York: St. Martin's Press, 1973.

Bakker, Robert. *The Dinosaur Heresies*. New York: William T. Morrow, 1986.

Barbour, Ian. *Issues in Science and Religion*. New York: Harper Torchbooks, 1966.

———. *Myths, Models and Paradigms*. San Francisco: Harper and Row, 1974.

———. *Religion and Science*. San Francisco: HarperSanFranciso, 1997.

———. *Religion in an Age of Science*. San Francisco: HarperSanFrancisco, 1990.

Barrow, John D., and Frank J. Tipler. *The Anthropic Cosmological Principle*. New York: Oxford Univ. Press, 1986.

Behe, Michael. *Darwin's Black Box*. New York: Free Press, 1986.

Bellah, Robert. *Beyond Belief*. New York: Harper and Row, 1970.

————. *Tokugawa Religion.* Glencoe, Ill.: The Free Press, 1957.

Bille, Matthew A. *Rumors of Existence.* Blaine, Wash.: Hancock House, 1995.

Booher, Harold. *Origins, Icons, and Illusions.* St. Louis: Warren H. Green, Inc. 1998.

Bridgman, Percy W. *The Way Things Are.* Cambridge, Mass.: Harvard Univ. Press, 1959.

————. *The Logic of Modern Physics.* New York: Macmillan, 1946.

Clube, Victor, and Bill Napier. *The Cosmic Serpent.* New York: Universe Press, 1982.

Collingwood, R. G. *The Idea of History.* New York: Galaxy Books, 1956.

Davies, Paul. *The Mind of God.* New York: Simon and Schuster, 1992.

de Grazia, Alfred. *The Lately Tortured Earth.* Princeton, N.J.: Metron Publications, 1983.

Densmore, Frances. "Teton Sioux Music," Bureau of American Ethnography, Bulletin 61, Smithsonian Institution, 1918.

Desmond, Adrian. *Hot-Blooded Dinosaurs.* New York: The Dial Press/James Wade, 1976.

Donnelly, Ignatius. *Ragnarok: The Age of Fire and Gravel.* New York: University Books, 1970.

Downey, Roger. *The Riddle of the Bones.* New York: Springler-Verlag New York. 2000.

Durrance, E. M. *Radioactivity in Geology.* Chester, England: Ellis Horwood, Ltd., 1986.

Eastman, Charles. *Indian Boyhood.* New York: McClure, Phillips & Co., 1902.

Eldredge, Niles. *Time Frames.* New York: Simon and Schuster, 1985.

Feyerabend, Paul. *Against Method.* New York: Schocken Books, 1975.

————. *Science in a Free Society.* New York: LNB, 1978.

Foster, David. *The Philosophical Scientists.* New York: Dorset Press, 1985.

Gilkey, Langdon. *Creationism on Trial.* Minneapolis: Winston Press, 1985.

Ginzberg, Louis. *The Legends of the Jews.* Philadelphia: Jewish Publishing Society of America, 1925.

Gould, Stephen Jay. *Ever Since Darwin.* New York: W. W. Norton & Co., 1977.

————. *The Mismeasure of Man.* New York: W. W. Norton & Co., 1981.

————. *Rocks of Ages.* New York: The Ballantine Publishing Groups, 1999.

Hadd, John. *Evolution: Reconciling the Controversy.* Glassboro, N.J.: Kronos Press, 1979.

Hall, Edward T. *Beyond Culture.* New York: Anchor-Doubleday, 1976.

Hays, H. R. *In the Beginning.* New York: Putnam, 1963.

Heisenberg, Werner. *Across the Frontiers.* New York: Harper and Row, 1976.

———. *The Physicist's Conception of Nature.* New York: Harcourt, Brace and Co., 1955.

———. *Physics and Philosophy.* New York: Harper Torchbooks, 1958.

Hooker, Dolph Earl. *Those Astounding Ice Ages.* New York: Exposition Press, 1958.

Hoyle, Sir Fred, and N. Chandra Wickramasinghe. *Evolution from Space.* New York: Simon and Schuster, 1981.

Johnson, Robert. *Darwin on Trial.* Washington, D.C.: Regnery Gateway, 1991.

Kitcher, Philip. *Abusing Science.* Cambridge, Mass.: MIT Press, 1982.

Kuhn, Thomas. *The Structure of Scientific Revolutions.* Chicago: University of Chicago Press, 1970.

Küng, Hans. *Eternal Life.* New York: Doubleday, 1984.

Küng, Hans. *Theology for the Third Millennium.* Oxford: One World Publications, 1988.

Lindley, David. *The End of Physics.* New York: Basic Books, 1993.

Lowie, Robert. *Primitive Religion.* New York: Liveright, 1948.

Macbeth, Norman. *Darwin Retried.* Cambridge, Mass.: Harvard Common Press, 1979).

Macquarrie, John. *Three Issues in Ethics.* New York: Harper and Row, 1970.

Milton, Richard. *Forbidden Science.* London: Fourth Estate, 1994.

———. *Shattering the Myths of Darwinism.* Rochester, Vt.: Park Street Press, 1997.

Mooney, Richard. *Gods of the Air and Darkness.* New York: Stein and Day, 1975.

Neville, Robert Cummings. *Behind the Masks of God.* Albany, N.Y.: State University of New York Press, 1991.

Niebuhr, H. Richard. *The Responsible Self.* New York: Harper and Row, 1963.

Pagels, Heinz. *Perfect Symmetry.* New York: Bantam Books, 1986.

Patten, Donald. *The Biblical Flood and the Ice Epoch.* Seattle: Pacific Meridian Publishing, 1966.

Patten, Donald, Ronald R. Hatch, and Loren Steinhaur. *The Long Day of Joshua and Six Other Catastrophes.* Seattle: Pacific Meridian Publishing, 1973.

Peat, David. *Synchronicity.* New York: Bantam Books, 1987.

Ring, Kenneth. *Life at Death.* New York: Coward, McCann and Geoghegan, 1980.

Rouse, W. D. *Catalysts for Change.* New York: John Wiley & Sons, 1993.

Santillana, Giorgio, and Hertha von Dechend. *Hamlet's Mill.* Boston: Gambit Books, 1969.

Schaer, Hans. *Religion and the Care of Souls in Jungian Psychology.* New York:

Bollingen Series XXI, Pantheon Books, 1950.

Sharpe, Eric J. *Nathan Söderblom and the Study of Religion.* Chapel Hill, N.C.: University of North Carolina Press, 1990.

Sheldrake, Rupert. *The Rebirth of Nature.* New York: Bantam Books, 1980.

Shklovskii, I. S., and Carl Sagan. *Intelligent Life in the Universe.* San Francisco: Holden-Day, 1966.

Simpson, George Gaylord. *The Major Features of Evolution.* New York: Simon and Schuster, 1953.

Sitchin, Zecharia. *The Twelfth Planet.* New York: Stein and Day, 1976.

Skinner, Brian J., and Stephen C. Porter. *The Dynamic Earth.* New York: John Wiley & Sons, 1992.

Slater, Philip. *The Wayward Gate.* Boston: Beacon Press, 1977.

Smith, Huston. *The Religions of Man.* New York, Harper and Row, 1958.

Söderblom, Nathan. *The Living God.* London: Oxford Univ. Press, 1933.

Somit, Albert, and Steven Peterson (eds.). *The Dynamics of Evolution.* Ithaca, N.Y.: Cornell Univ. Press, 1992.

Stanley, Steven. *The New Evolutionary Timetable.* New York: Basic Books, 1981.

Supplement to the Vetus Testament, vol. II. Leiden, Netherlands: E. J. Brill, 1955, pp. 18–19.

INDEX